# Optics at the Nanometer Scale

# NATO ASI Series

## Advanced Science Institutes Series

*A Series presenting the results of activities sponsored by the NATO Science Committee, which aims at the dissemination of advanced scientific and technological knowledge, with a view to strengthening links between scientific communities.*

The Series is published by an international board of publishers in conjunction with the NATO Scientific Affairs Division

| | |
|---|---|
| **A  Life Sciences** | Plenum Publishing Corporation |
| **B  Physics** | London and New York |
| | |
| **C  Mathematical and Physical Sciences** | Kluwer Academic Publishers |
| **D  Behavioural and Social Sciences** | Dordrecht, Boston and London |
| **E  Applied Sciences** | |
| | |
| **F  Computer and Systems Sciences** | Springer-Verlag |
| **G  Ecological Sciences** | Berlin, Heidelberg, New York, London, |
| **H  Cell Biology** | Paris and Tokyo |
| **I  Global Environmental Change** | |

### PARTNERSHIP SUB-SERIES

| | |
|---|---|
| 1. **Disarmament Technologies** | Kluwer Academic Publishers |
| 2. **Environment** | Springer-Verlag / Kluwer Academic Publishers |
| 3. **High Technology** | Kluwer Academic Publishers |
| 4. **Science and Technology Policy** | Kluwer Academic Publishers |
| 5. **Computer Networking** | Kluwer Academic Publishers |

*The Partnership Sub-Series incorporates activities undertaken in collaboration with NATO's Cooperation Partners, the countries of the CIS and Central and Eastern Europe, in Priority Areas of concern to those countries.*

### NATO-PCO-DATA BASE

The electronic index to the NATO ASI Series provides full bibliographical references (with keywords and/or abstracts) to more than 50000 contributions from international scientists published in all sections of the NATO ASI Series.
Access to the NATO-PCO-DATA BASE is possible in two ways:

– via online FILE 128 (NATO-PCO-DATA BASE) hosted by ESRIN,
Via Galileo Galilei, I-00044 Frascati, Italy.

– via CD-ROM "NATO-PCO-DATA BASE" with user-friendly retrieval software in English, French and German (© WTV GmbH and DATAWARE Technologies Inc. 1989).

The CD-ROM can be ordered through any member of the Board of Publishers or through NATO-PCO, Overijse, Belgium.

**Series E: Applied Sciences - Vol. 319**

# Optics at the Nanometer Scale

## Imaging and Storing with Photonic Near Fields

edited by

## M. Nieto-Vesperinas

Instituto de Ciencia de Materiales,
Consejo Superior de Investigaciones Científicas,
Cantoblanco, Madrid, Spain

and

## N. García

Laboratorio de Física de Sistemas Pequeños,
Consejo Superior de Investigaciones Científicas,
Cantoblanco, Madrid, Spain

**Kluwer Academic Publishers**

Dordrecht / Boston / London

Published in cooperation with NATO Scientific Affairs Division

Proceedings of the NATO Advanced Research Workshop on
Near Field Optics: Recent Progress and Perspectives
Miraflores, Madrid, Spain
September 11–15, 1995

A C.I.P. Catalogue record for this book is available from the Library of Congress.

ISBN 0-7923-4020-5

Published by Kluwer Academic Publishers,
P.O. Box 17, 3300 AA Dordrecht, The Netherlands.

Kluwer Academic Publishers incorporates the publishing programmes of
D. Reidel, Martinus Nijhoff, Dr W. Junk and MTP Press.

Sold and distributed in the U.S.A. and Canada
by Kluwer Academic Publishers,
101 Philip Drive, Norwell, MA 02061, U.S.A.

In all other countries, sold and distributed
by Kluwer Academic Publishers Group,
P.O. Box 322, 3300 AH Dordrecht, The Netherlands.

*Printed on acid-free paper*

Printed in the Netherlands

This book contains the proceedings of a NATO Advanced Research Workshop held within the programme of activities of the NATO Special Programme on Nanoscale Science as part of the activities of the NATO Science Committee.

Other books previously published as a result of the activities of the Special Programme are:

NASTASI, M., PARKING, D.M. and GLEITER, H. (eds.), *Mechanical Properties and Deformation Behavior of Materials Having Ultra-Fine Microstructures.* (E233) 1993 ISBN 0-7923-2195-2

VU THIEN BINH, GARCIA, N. and DRANSFELD, K. (eds.), *Nanosources and Manipulation of Atoms under High Fields and Temperatures: Applications.* (E235) 1993 ISBN 0-7923-2266-5

LEBURTON, J.-P., PASCUAL, J. and SOTOMAYOR TORRES, C. (eds.), *Phonons in Semiconductor Nanostructures.* (E236) 1993 ISBN 0-7923-2277-0

AVOURIS, P. (ed.), *Atomic and Nanometer-Scale Modification of Materials: Fundamentals and Applications.* (E239) 1993 ISBN 0-7923-2334-3

BLÖCHL, P. E., JOACHIM, C. and FISHER, A. J. (eds.), *Computations for the Nano-Scale.* (E240) 1993 ISBN 0-7923-2360-2

POHL, D. W. and COURJON, D. (eds.), *Near Field Optics.* (E242) 1993 ISBN 0-7923-2394-7

SALEMINK, H. W. M. and PASHLEY, M. D. (eds.), *Semiconductor Interfaces at the Sub-Nanometer Scale.* (E243) 1993 ISBN 0-7923-2397-1

BENSAHEL, D. C., CANHAM, L. T. and OSSICINI, S. (eds.), *Optical Properties of Low Dimensional Silicon Structures.* (E244) 1993 ISBN 0-7923-2446-3

HERNANDO, A. (ed.), *Nanomagnetism* (E247) 1993. ISBN 0-7923-2485-4

LOCKWOOD, D.J. and PINCZUK, A. (eds.), *Optical Phenomena in Semiconductor Structures of Reduced Dimensions* (E248) 1993. ISBN 0-7923-2512-5

GENTILI, M., GIOVANNELLA, C. and SELCI, S. (eds.), *Nanolithography: A Borderland Between STM, EB, IB, and X-Ray Lithographies* (E264) 1994. ISBN 0-7923-2794-2

GÜNTHERODT, H.-J., ANSELMETTI, D. and MEYER, E. (eds.), *Forces in Scanning Probe Methods* (E286) 1995. ISBN 0-7923-3406-X

GEWIRTH, A.A. and SIEGENTHALER, H. (eds.), *Nanoscale Probes of the Solid/Liquid Interface* (E288) 1995. ISBN 0-7923-3454-X

CERDEIRA, H.A., KRAMER, B. and SCHÖN, G. (eds.), *Quantum Dynamics of Submicron Structures* (E291) 1995. ISBN 0-7923-3469-8

WELLAND, M.E. and GIMZEWSKI, J.K. (eds.), *Ultimate Limits of Fabrication and Measurement* (E292) 1995. ISBN 0-7923-3504-X

EBERL, K., PETROFF, P.M. and DEMEESTER, P. (eds.), *Low Dimensional Structures Prepared by Epitaxial Growth or Regrowth on Patterned Substrates* (E298) 1995. ISBN 0-7923-3679-8

MARTI, O. and MÖLLER, R. (eds.), *Photons and Local Probes* (E300) 1995. ISBN 0-7923-3709-3

GUNTHER, L. and BARBARA, B. (eds.), *Quantum Tunneling of Magnetization - QTM '94* (E301) 1995. ISBN 0-7923-3775-1

PERSSON, B.N.J. and TOSATTI, E. (eds.), *Physics of Sliding Friction* (E311) 1996. ISBN 0-7923-3935-5

MARTIN, T.P. (ed.), *Large Clusters of Atoms and Molecules* (E313) 1996. ISBN 0-7923-3937-1

DUCLOY, M. and BLOCH, D. (eds.), *Quantum Optics of Confined Systems* (E314). 1996. ISBN 0-7923-3974-6

ANDREONI, W. (ed.), *The Chemical Physics of Fullereness 10 (and 5) Years Later. The Far-Reaching Impact of the Discovery of $C_{60}$* (E316). 1996. ISBN 0-7923-4000-0

# TABLE OF CONTENTS

Preface . . . . . . . . . . . . . . . . . . . . . . . . . . . . . . . . . . . . . . . . . . . . . . ix

**Theory and Basic Principles**

Theory of Imaging in Near-Field Microscopy
*J.J. Greffet and R. Carminati* . . . . . . . . . . . . . . . . . . . . . . . . . . . . . . . . 1

Light Scattering by Tips in Front of Surfaces
*A. Madrazo, M. Nieto-Vesperinas and N. García* . . . . . . . . . . . . . . . . . . . . 27

A Numerical Study of a Model Near-Field Optical Microscope
*A.A. Maradudin, A. Mendoza-Suárez, E.R. Méndez and M. Nieto-Vesperinas* . . . . . 41

Short and Long Range Interactions in Near Field Optics
*O. Keller* . . . . . . . . . . . . . . . . . . . . . . . . . . . . . . . . . . . . . . . . . . . . 63

Modelling Optical Resonators Probed by Subwavelength Sized Optical Detectors
*A. Castiaux, Ch. Girard, A. Dereux, X. Boujou and J.P. Vigneron* . . . . . . . . . . 95

**Experiments: Fundamentals and Applications**

Instrumentation in Near Field Optics
*D. Courjon, F. Baida, C. Bainier and D. Van Labeke* . . . . . . . . . . . . . . . . . . 105

Effect of the Coherence in Near Field Microscopy
*F. de Fornel, L. Salomon, J.C. Weeber, A. Rahmani, C. Pic and A. Dazzi* . . . . . . 119

Scanning Interferometric Apertureless Microscopy at Ten Angstrom Resolution
*H.K. Wickramasinghe, Y. Martin and F. Zenhausern* . . . . . . . . . . . . . . . . . . 131

Primary Imaging Modes in Near-Field Microscopy
*M. Vaez-Iravani* . . . . . . . . . . . . . . . . . . . . . . . . . . . . . . . . . . . . . . . . 143

Local Excitation of Surface Plasmons by TNOM
*B. Hecht, D.W. Pohl and L. Novotny* . . . . . . . . . . . . . . . . . . . . . . . . . . . 151

Weak Localization of Surface Plasmon Polaritons: Direct Observation with

Photon Scanning Tunneling Microscope
*S.I. Bozhevolnyi, A.V. Zayats and B. Vohnsen* . . . . . . . . . . . . . . . . . . . . . . 163

STM-Induced Photon Emission from Au (110)
*R. Berndt* . . . . . . . . . . . . . . . . . . . . . . . . . . . . . . . . . . 175

Writing of Nanolines on a Ferroelectric Surface with a Scanning Near-Field
Optical Microscope
*J. Massanell, N. García, A. Correia, A. Zlatkin, M. Sharonov and*
*J. Przeslawski* . . . . . . . . . . . . . . . . . . . . . . . . . . . . . . . . . 181

Near Field Optics with High-Q Whispering-Gallery Modes
*N. Dubreuil, J.C. Knight, J. Hare, V. Lefevre-Sequin, J.M. Raimond and*
*S. Haroche* . . . . . . . . . . . . . . . . . . . . . . . . . . . . . . . . . . 191

Fluorescence Microscopy and Spectroscopy by Scanning Near-Field Optical /
Atomic Force Microscope (SNOM-AFM)
*M. Fujihira* . . . . . . . . . . . . . . . . . . . . . . . . . . . . . . . . . . 205

Fluorescence Lifetime Contrast Combined with Probe Microscopy
*O.H. Willemsen, O.F.J. Noordman, F.B. Segerink, A.G.T. Ruiter, M.H.P. Moers*
*and N.F. Van Hulst* . . . . . . . . . . . . . . . . . . . . . . . . . . . . . . . 223

Towards SNIM: Scanning Near-Field Microscopy in the Infrared
*F. Keilmann* . . . . . . . . . . . . . . . . . . . . . . . . . . . . . . . . . . 235

6 NM Lateral Resolution in Scanning Near Field Optical Microscopy with the
Tetrahedral Tip
*J. Koglin, U.C. Fischer and H. Fuchs* . . . . . . . . . . . . . . . . . . . . . . . 247

An Aperture-Type Reflection-Mode SNOM
*C. Durkan and I.V. Shvets* . . . . . . . . . . . . . . . . . . . . . . . . . . . . 257

Surface Modifications Via Photo-Chemistry in a Reflection Scanning Near-Field
Optical Microscope
*S. Madsen, T. Olesen and J.M. Hvam* . . . . . . . . . . . . . . . . . . . . . . . 263

Near-Field Diffraction Microscopy with a Coherent Low-Energy e-Beam:
Fresnel Projection Microscope
*Vu Thien Binh, V. Semet, N. García and L. Bitar* . . . . . . . . . . . . . . . . . . 277

Author Index . . . . . . . . . . . . . . . . . . . . . . . . . . . . . . . . . . 297

Subject Index . . . . . . . . . . . . . . . . . . . . . . . . . . . . . . . . . . 299

# PREFACE

This book contains most of the lectures presented at the NATO Advanced Research Workshop on Near Field Optics: Recent Progress and Perspectives, held at La Cristalera, Miraflores, Madrid, from 11 through September 15, 1995. More than thirty-five scientists from Europe, the United States and Japan attended this meeting and contributed with live talks and discussions about their recent research. Eleven sessions with a total of twenty-seven talks were given. These included topics ranging from the fundamentals of near field optical microscopy imaging and new interferometric methods, to applications in fluorescence and information encoding.

The workshop reflected the impressive progress made in the understanding, development and applications of this subject, it also established paths for future work that may be undertaken in the very near future: the continuous improvement of tips and illumination-collection configurations, which will permit researchers to penetrate atomic resolution, both for observing and storing, and the setting up of controlled experiments and increasingly refined theoretical models that will permit a matching together of the results in this new and fascinating science.

The workshop was organized under the auspices and financial support of NATO. Additional help was also provided by the Comisión Interministerial de Ciencia y Tecnología, the Consejo Superior de Investigaciones Científicas, the Comunidad de Madrid, and the Universidad Autónoma de Madrid.

We thank all the authors for their contributions, and the afore-mentioned institutions for their support. Special thanks are given to the NATO Science Committee who made this workshop possible.

The editors

# THEORY OF IMAGING IN NEAR-FIELD MICROSCOPY

J.-J. GREFFET and R. CARMINATI
Laboratoire d'Energétique Moléculaire et Macroscopique; Combustion
Ecole Centrale Paris, CNRS
F-92295 Châtenay-Malabry Cedex, France

## 1. Introduction

In the last ten years, several different set-ups for producing images with super resolution by working in the near field have been reported. A good overwiew of the subject can be found in some recent review papers [1-3]. The techniques can be split roughly into two categories. Super-resolution may be obtained by detecting in the far field the light scattered by a small part of the sample under very localized illumination (illumination mode Scanning Near-field Optical Microscope), or, alternatively, by detecting the near field very close to the surface illuminated from the far field (detection mode SNOM). The most efficient way to produce a localized illumination is the use of coated tapered fibers first introduced by Betzig et al. [4] (see Fig. 1a). Localized detection can be achieved by bringing the narrow tip of an optical fiber close to the surface and detecting the signal coupled into the fiber (see Fig. 1b). Alternatively, a small scatterer may be introduced in the near field so that the scattered light can be detected in the far field (see Fig. 1c). In this paper, we shall discuss the second type of experiments. We shall also restrict the discussion to elastic scattering, thus excluding fluorescence.

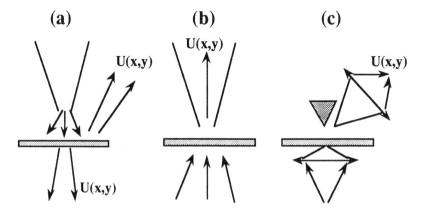

*Figure 1.* Experimental set-ups used in near-field microscopy

1

M. Nieto-Vesperinas and N. García (eds.), Optics at the Nanometer Scale 1–26.
© 1996 Kluwer Academic Publishers. Printed in the Netherlands.

As for other scanning techniques, the key issue is to develop a simple relationship between the images and the structure of the sample. Consider for instance the set-up that uses the tip of an optical fiber scanned above the sample. The output is the signal delivered by the fiber to the detector for a given position of the tip along the sample. A series of scans at discrete positions yields an image. The problem is to understand the relationship between this image and the actual structure of the sample. Before achieving this task, it is necessary to identify and understand the mechanisms that produce a contrast in the image. To some extent, this problem is essentially a scattering problem. Given the structure of the sample and the illumination conditions, is it possible to determine the near field ? Yet, the problem of image formation is more difficult since it involves the coupling of the near field with the fiber modes.

In order to interpret these images, a simple experimental solution would be to look for specific signatures of some features of a given sample. Assume that under a given illumination, with a specified angle of incidence and polarization, a decay of the dielectric contrast leads to a decay of the signal. Is there any reason to believe that such a signature is not ambiguous ? We shall show in this paper that a modification of the topopraphy could lead to the same variation of the signal. The problem is then to identify the quantity that is being imaged. We may expect this quantity to depend on topography, dielectric contrast and on illumination conditions. Another important issue is whether the relation between the sample and the image is local. If the signal detected at a given point is related to the topography at the same point, then it turns out that the contribution of the topography can be substracted. This locality is ensured in localized illumination SNOM. However, in localized detection SNOM, the relation is usually a non-local relation. We will show that under special illumination conditions, this locality can be restored to some extent. An alternative solution is to process the image to retrieve the actual surface profile [5-8]. This goal appears to be very difficult at first glance due to the multiple scattering between the tip and the sample. In some cases, it appears possible to neglect this multiple interaction. Thus, one might expect the signal delivered by the detector to be proportional to the near field illuminating the tip. This amounts to considering the tip a passive probe. If this assumption is valid, then the analysis of the formation of the image is considerably simplified and can be split into two steps (see Fig.2 a). The first step is the description of the near field scattered by the sample in the absence of the tip. This is a standard scattering problem. The second step is the description of the coupling of the scattered intensity $I(x,y)$ with the optical fiber which delivers an output signal $U(x,y)$.

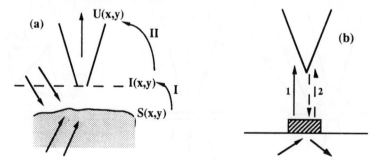

*Figure 2*. A schematic diagram of the imaging process.

In this paper we shall discuss the modelling of these two steps. The following section is devoted to the discussion of the passive probe assumption. In section 3, we will introduce a perturbative framework to analyse the scattering of the field in the near-field region. This formalism is useful for discussing the image contrast mechanism and in analysing the coupling of the dielectric contrast and topography in the image formation process. We will show that this formalism leads in a natural way to the introduction of a transfer function between the sample properties and the intensity of the field. This concept will be used to discuss the imaging properties of the usual Photon Scanning Tunneling Microscope (PSTM) and how it can be improved. It will be shown that the transfer function is appropriate for discussing the effects of the polarization and coherence of the illuminating field. Finally, we will show how this concept can be used to solve the inverse problem, i.e. how it is possible to retrieve the structure of the sample from the near-field intensity. The last section is devoted to the analysis of the speckle pattern in the near field produced by the residual random roughness of any real surface.

## 2. The assumption of a passive probe

The passive probe assumption considers that the probe is illuminated by the near field that would exist if the tip was absent. Thus, the assumption is correct if the field scattered by the tip towards the sample and scattered back towards the tip can be neglected (See path 2 in Fig.2 b). This hypothesis may be valid either if the field scattered by the tip is very small or if it is not reflected back by the sample. Thus, from this qualitative analysis, we may expect the probe to be passive either if the tip is very small or if the sample has a low reflectivity. Therefore, a metallic tip close to a metallic sample may not satisfy the assumption of a passive probe whereas a tiny metallic tip above a dielectric surface or a dielectric tip above a metal surface might be considered as passive probes. Note that the passive probe assumption may be valid even if the field close to the tip is strongly perturbed. As an example consider a tip which is modelled by a small sphere whose dielectric constant ($\varepsilon_{int}$) is substantially different from that of the surrounding medium ($\varepsilon_{ext}$). If the radius of the sphere is much smaller than the wavelength, then the electrostatic approximation is valid and we know that the field inside the sphere $\mathbf{E}_{int}$ is related to the external field $\mathbf{E}_{ext}$ by the relationship

$$\mathbf{E}_{int} = \frac{3\varepsilon_{ext}}{2\varepsilon_{ext} + \varepsilon_{int}} \mathbf{E}_{ext}$$

(1)

The field inside the sphere is proportional to the field that would exist if the sphere was absent, which is necessary for the passive probe assumption. But the near field close to the sphere is strongly affected by its presence, due to the fact that the field has to satisfy the continuity relations.

Some results are available in the literature to confirm this analysis. Totzeck and Krumbügel [9, 10] have reported a comparison between experiment and theory in the microwave domain. They have measured the amplitude and phase of the field scattered by a rectangular strip for three different dielectric materials (indices n=1.018+i0.0002, n=1.43+i0.001 and n=4.1) using a metallic waveguide as a probe for TE polarization. The probe was located at a distance of 0.09 λ above the surface. The agreement with the theory is excellent *although the tip was not taken into account in the computations*. This example is very interesting since the situation is at first sight quite demanding for the passive probe assumption. The tip is metallic, its width is 0.09 λ, it is located at 0.09 λ from the surface and the sample material has a relatively large index (n=4.1). However, the signal detected remains proportional to the scattered field computed *in the absence of the tip*.

As a second example, we quote a comparison between PSTM images of a quartz step and numerical calculations done without accounting for the presence of the tip [11]. In this case, both the tip and the surface have an index of 1.5. A good qualitative agreement was found verifying the assumption of a passive probe. Note that we have compared the experimental signal with the square modulus of the electric field (intensity). It was by no means obvious that the relevant quantity was the intensity.

Finally, in order to investigate further the interaction between the tip and sample, we have developed a complete numerical simulation of a two-dimensional PSTM set-up [12]. The system studied is depicted in Fig.3 a. The distance x between the ridge and the tip is varied along a line at constant height $z_0$. For each position x of the tip, a calculation accounting for the multiple scattering between the tip and sample is done and the flux of the Poynting vector in the fiber is computed. This flux is the signal U(x) that is compared with the intensity I(x) defined as the square modulus of the electric field at the tip vertex. Note that we consider the intensity computed in the absence of the tip. A transfer function can be defined between the signal and the intensity by introducing the ratio of their Fourier transforms U(k)/I(k). Fig.3 b shows the modulus of the transfer function when scanning different surface profiles with the same probe. We observe that the transfer function obtained is essentially the same. This indicates that this quantity depends only on the fiber and does not depend on the surface itself as it would if multiple scattering could not be neglected.

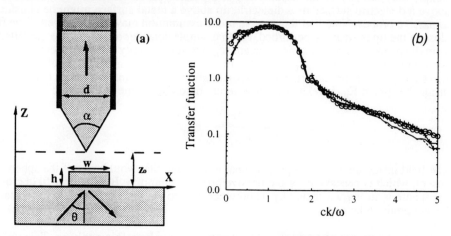

*Figure 3*. Numerical simulation of a PSTM. (a) set-up (b) transfer function U(k)/I(k).

The consequence of this result is important for the analysis of the image formation : we can split the analysis of the process into two steps. The generation of the signal through the coupling of the near-field intensity with the modes of the optical fiber can be simply accounted for by using a transfer function (step II in Fig.2 a). Thus, the remaining problem is to calculate the field scattered by the sample (step I). This can be done through numerical simulation. Yet, in order to gain physical insight into the scattering process, it is desirable to develop a simplified analytical approach of the problem. This is the goal of the next section.

## 3. Perturbative analysis of the scattered field

In this section, we introduce the basic concepts necessary for a discussion of the scattering process. From section 2, we know that the relevant quantity to be calculated is the intensity defined as the square modulus of the electric field. We also know that we do not need to take the perturbation of the tip into account. We shall try to find the signature of the sample features which is present in the near-field intensity. Only the main lines of the theoretical development will be presented here. The reader is refered to [13] for further details.

### 3.1 GEOMETRY OF THE SYSTEM

The system considered in this section is depicted in Fig.4. We assume that the medium is homogeneous in the half space $\Omega_3$ defined by $z<0$ with a complex scalar dielectric constant $\varepsilon_3$. The upper half space $\Omega_1$ is defined by the condition $z>S(x,y)$ and is characterized by the complex scalar dielectric constant $\varepsilon_1$. The intermediate region $\Omega_2$ $0<z<S(x,y)$ may be inhomogeneous and is described by the dielectric constant $\varepsilon_2(x,y,z)$. This structure scatters light due to the dielectric constant variations and the topography. The system may be illuminated either from below or from above and we are interested in the scattered field above the sample ($z>S(x,y)$). The analysis is performed in the framework of classical electrodynamics. We assume that the dielectric constant is a meaningful quantity even for mesoscopic systems. This problem has been discussed by Keller [14-16].

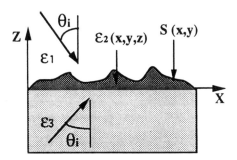

*Figure 4* . Geometry of the system.

## 3.2 INTEGRAL EQUATION

From Maxwell's equations, an integral equation for the electric field can be derived (a temporal dependence $\exp(-i\omega t)$ is assumed for all fields) :

$$\mathbf{E}(\mathbf{r}) = \mathbf{E}_f(\mathbf{r}) + \int_{\Omega_2} k_0^2 [\varepsilon(\mathbf{r}') - \varepsilon_1] \overline{\overline{\mathbf{G}}}(\mathbf{r}_{\parallel} - \mathbf{r}_{\parallel}', z, z') \bullet \mathbf{E}(\mathbf{r}') d\mathbf{r}' \qquad (2)$$

where $\mathbf{r} = (\mathbf{r}_{\parallel}, z)$ is a point of $\Omega_1$ and $k_0 = \omega/c$.

$\overline{\overline{\mathbf{G}}}(\mathbf{r}_{\parallel} - \mathbf{r}_{\parallel}', z, z')$ is the Green dyadic that accounts for the presence of the lower medium, $\mathbf{E}_f$ is the field that would exist for a flat interface separating $\Omega_1$ and $\Omega_3$. This is either a transmitted field or the sum of the incident and reflected fields. The integral term is the field scattered by both the topography and the inhomogeneities of $\Omega_2$. The term $k_0^2 [\varepsilon(\mathbf{r}') - \varepsilon_1] \mathbf{E}(\mathbf{r}') d\mathbf{r}'$ can be interpreted as the induced dipole of the elementary volume $d\mathbf{r}'$. The dot product of the dipole located in $\mathbf{r}'$ with the Green's dyadic $\overline{\overline{\mathbf{G}}}(\mathbf{r}_{\parallel} - \mathbf{r}_{\parallel}', z, z')$ (also called propagator) gives the expression of the electric field scattered towards $\mathbf{r}$. Note that the integral is performed only over the domain $\Omega_2$ of the scatterer. Thus, the scattered field is known everywhere as soon as it is known in $\Omega_2$. To this aim, one can restrict Eq.(2) to $\Omega_2$ and numerically solve the integral equation. The total field can then be calculated using (2) everywhere, providing a reference solution that we will use to test our approximate solutions [17-19]. Girard has developped a microscopic model based on the use of a self-consistent equation [20, 21]. The equivalence between the macroscopic approach (Eq.(2)) and the self-consistent microscopic approach is derived in ref.[22].

## 3.3 EQUIVALENT SURFACE PROFILE

A simple expression of the scattered field can be obtained using the Born approximation. A full analysis of the domain of validity of this approximation has been reported elsewhere [13]. It amounts to replacing the electric field in the integral in Eq.(2) by $\mathbf{E}_f$. Note that this is equivalent to using the first term in the iterative solution of the integral equation. Thus, we will use the notation $\mathbf{E}^{(0)}$ for $\mathbf{E}_f$ and $\mathbf{E}^{(1)}$ for the scattered field under the Born approximation. The second approximation neglects the $z'$ dependence of the zero-order electric field in the integral descibing $\mathbf{E}^{(1)}$. This is meaningful because in near-field optics, the maximum value of $S(x,y)$ is on the order of tens of nanometers. Thus, the phase and amplitude variations of an electric field whose wavelength is on the order of a micrometer can be neglected. For the same reason, the $z'$ dependence of the Green dyadic can be neglected. This assumption is discussed in several references in the context of scattering by rough surfaces [23-25]. It breaks down in the case of an illumination with a coated fiber since the amplitude of the illuminating field varies rapidly. Thus this analysis is restricted to far-field illumination. The two assumptions can be summarized by the equation :

$$\overline{\overline{\mathbf{G}}}(\mathbf{r}_{\parallel} - \mathbf{r}_{\parallel}', z, z') \bullet \mathbf{E}(\mathbf{r}_{\parallel}', z') \approx \overline{\overline{\mathbf{G}}}(\mathbf{r}_{\parallel} - \mathbf{r}_{\parallel}', z, 0) \bullet \mathbf{E}_f(\mathbf{r}_{\parallel}', 0) \qquad (3)$$

Substituting this expression in Eq.(2) and rearranging terms, one obtains the following expression for the scattered field

$$\mathbf{E}^{(1)}(\mathbf{r}_\parallel, z) = \int_{\Omega_2} d\mathbf{r}_\parallel' \, k_0^2 \overline{\overline{\mathbf{G}}}(\mathbf{r}_\parallel - \mathbf{r}_\parallel', z, z') \bullet \mathbf{E}^{(0)}(\mathbf{r}_\parallel', 0) \int_0^{S(\mathbf{r}_\parallel')} \left[ \varepsilon(\mathbf{r}_\parallel', z') - \varepsilon_1 \right] dz' \qquad (4)$$

Eq.(4) shows that all the information on the scatterer appears in the coupled term

$$\int_0^{S(\mathbf{r}_\parallel')} \left[ \varepsilon(\mathbf{r}_\parallel', z') - \varepsilon_1 \right] dz' \qquad (5)$$

This has an important consequence. Two different samples with different topography and different dielectric contrasts but the same value of the integral in Eq.(5) *produce the same near field*. Therefore, it is impossible to discriminate between them. A new concept has to be introduced to deal with this property. We can either define an equivalent surface profile or an equivalent dielectric contrast. We have chosen to define an equivalent surface profile :

$$S_{eq}(\mathbf{r}_\parallel') = \frac{1}{\varepsilon_3 - \varepsilon_1} \int_0^{S(\mathbf{r}_\parallel')} \left[ \varepsilon(\mathbf{r}_\parallel', z') - \varepsilon_1 \right] dz' \qquad (6)$$

This structure of the scattered field shows that a single image is not sufficient to determine the sample. It is necessary to obtain the topography by other means or to perform somehow a tomography. The question remains open. Note that we could also have defined an equivalent surface current on the sample surface, as was done by Kröger and Kretschmann in the context of scattering by rough surfaces [26].

In the derivation of Eq.(5), we have made some approximations. Therefore, it is important to check that the predictions based on Eq.(5) are accurate for the kind of samples that are used in near-field microscopy. To this end, we have used a complete two-dimensional numerical simulation to compute the near-field intensity for different structures that should produce the same near field according to Eq.(5). The domain $\Omega_2$ is a ridge having a width $\lambda/2$, a dielectric constant $\varepsilon_2$ and a height h such that $\Delta\varepsilon$ h remains constant, where $\Delta\varepsilon = (\varepsilon_2 - \varepsilon_1)$ is the dielectric contrast. The lower medium $\Omega_3$ has a dielectric constant $\varepsilon_3 = 2.25$, the upper medium $\Omega_1$ is a vacuum ($\varepsilon_1 = 1$). We represent the intensity in $\Omega_1$ along a line parallel to the interface at a distance $z_0 = 70$ nm above the surface. The system is illuminated in transmission from $\Omega_3$ by a plane wave of wavelength $\lambda = 633$ nm. Fig.5 a-b show the intensity for a ridge deposited on the interface (surface defect) for s and p-polarization respectively, for normal incidence. It can be seen that the agreement is better than 10%, confirming the validity of the perturbative analysis. The approximation is better for s-polarization. This is due to the fact that the Born approximation works better for s-polarization [13] than for p-polarization. The intensity along the line $z_0 = 40$ nm for a ridge burried in the lower medium (subsurface defect) is shown in Fig.5 c for an angle of incidence of 45° (i.e. illumination in total internal reflection) in p-polarization. Note that in this case the dielectric contrast $\Delta\varepsilon$ is defined by ($\varepsilon_2 - \varepsilon_3$).

8

In Fig.5 d we compare a surface defect with a subsurface defect. The analysis is slightly different in this case because the exact form of the Green dyadic is not the same for a source point located below or above the interface. Thus we cannot derive a general equivalence rule for dips and bumps. However, the figure shows that the bump and the dip may produce the same field. Thus an image cannot tell us if we have a dip or a bump. The above results can be summarized in a simple way : *a simple correlation between the contrast of the image and the structure of the sample does not exist*. A systematic study, either experimental or numerical, trying to correlate the contrast of the image with some sample features cannot produce any general rule of thumb.

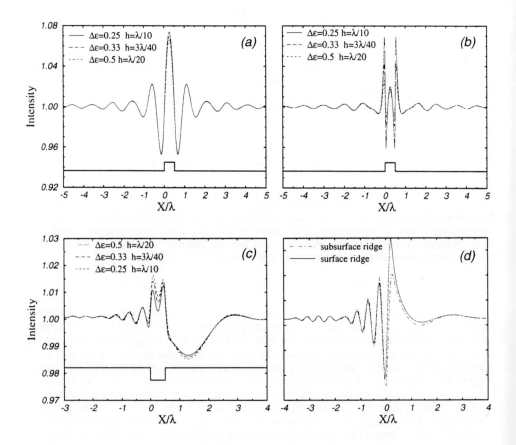

*Figure 5*. Comparison of near-field intensities produced by different subwavelength structures with the same equivalent surface profile.

## 3.4 PERTURBATIVE EXPRESSION OF THE FIELD SCATTERED BY A ROUGH SURFACE

In section 3.3 we have shown, under some conditions usually satisfied by the samples used in near-field microscopy, that it is possible to define an equivalent surface profile that describes the scattering by any inhomogeneous system. Therefore, it suffices to solve the problem of scattering by a homogeneous rough surface defined by its profile S(x,y). In the following, one has to keep in mind that for an inhomogeneous sample, S(x,y) denotes the *equivalent* surface profile. The system studied in this section is depicted in Fig.6. Two homogeneous media (dielectric constant $\varepsilon_1$ and $\varepsilon_2$) are separated by a rough interface, defined by its profile S(x,y). For the sake of clarity, we restrict our analysis to an illumination in transmission from the lower medium, and to an evaluation of the near field in the upper region.

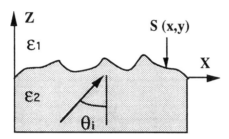

*Figure 6.* Scattering geometry.

We introduce an expression of the scattered field in Fourier space. This proves to be very useful for a discussion of the super-resolution issue in terms of spatial frequencies. Let us start by introducing the Fourier transform of the electric field

$$\mathbf{E}(\mathbf{r}) = \int d\mathbf{k}_{\parallel} \; \mathbf{e}(\mathbf{k}_{\parallel}) \exp[\, i\,(\mathbf{k}_{\parallel}.\,\mathbf{r}_{\parallel} + \gamma_1 z)], \qquad (7)$$

where $\gamma_1(\mathbf{k}_{\parallel}) = [\varepsilon_1 \omega^2/c^2 - \mathbf{k}_{\parallel}^2]^{1/2}$ with the condition $\mathrm{Im}(\gamma_1) > 0$, $\mathrm{Re}(\gamma_1) > 0$. This representation is valid at any point located above the highest point of the surface. Let us define the Fourier transform of the surface profile by

$$S(\mathbf{k}_{\parallel}) = \int d\mathbf{r}_{\parallel} \; S(\mathbf{r}_{\parallel}) \exp[\, i\,(\mathbf{k}_{\parallel}.\mathbf{r}_{\parallel})] \qquad (8)$$

The amplitude of each plane wave can be written as in section 3.3 as the sum of a zero order term (which corresponds to the case of a flat interface) and a first order term due to scattering.

$$\mathbf{e}(\mathbf{k}_{\parallel}) = \mathbf{e}^{(0)}(\mathbf{k}_{\parallel}) + \mathbf{e}^{(1)}(\mathbf{k}_{\parallel}) \qquad (9)$$

There are a number of different approaches to solving the problem of scattering by rough surfaces to the first order [23-25, 27-30]. To obtain the detailed expression of the amplitudes of the field in Fourier space, one can insert the Fourier transform of the Green dyadic Eq.(7-8) in Eq.(4) [13]. In the case of an illumination by a plane wave in transmission with wave vector $\mathbf{k}_{inc}$ and amplitude $\mathbf{e}_{inc}$, the general form of the field is given by

$$\mathbf{e}^{(0)}(\mathbf{k}_{\parallel}) = \left[t\left(\mathbf{k}_{\parallel}^{inc}\right)\right]\mathbf{e}_{inc}\,\delta\left(\mathbf{k}_{\parallel} - \mathbf{k}_{\parallel}^{inc}\right)$$

$$\mathbf{e}^{(1)}(\mathbf{k}_{\parallel}) = S\left(\mathbf{k}_{\parallel}^{inc} - \mathbf{k}_{\parallel}\right)L\left(\mathbf{k}_{\parallel},\mathbf{k}_{\parallel}^{inc}\right)\mathbf{e}_{inc} \tag{10}$$

where $\left[t\left(\mathbf{k}_{\parallel}^{inc}\right)\right]$ is a matrix of transmission factors. L is a linear operator given in refs [11, 31]. Note that the spectrum of the scattered field is linearly related to the spectrum of the surface. Yet, it has been shifted by $\mathbf{k}_{\parallel}^{inc}$. The physical meaning of this result is clear in the case of a slightly rough surface. The diffuse field is concentrated around the specular direction. The operator L can be represented by a 2X2 matrix if one represents the field using the s and p components or by a 3X3 matrix if the field is represented by its cartesian components. The important point to note is that the matrix elements of L are analytically known. The scattered field can be cast in the form :

$$\mathbf{E}^{(1)}(\mathbf{r}) = \int d\mathbf{k}_{\parallel}\, S\left(\mathbf{k}_{\parallel}^{inc} - \mathbf{k}_{\parallel}\right)L\left(\mathbf{k}_{\parallel},\mathbf{k}_{\parallel}^{inc}\right)\mathbf{e}_{inc}\,\exp\left[i(\mathbf{k}_{\parallel}\cdot\mathbf{r}_{\parallel} + \gamma_1(\mathbf{k}_{\parallel})z)\right] \tag{11}$$

This expression is equivalent to Eq.(4), S being the equivalent surface profile. Besides the shift in spatial frequency, another salient feature appears in this expression. As it has been discussed in many papers [11, 29, 30, 32, 33], the contribution from the high spatial frequencies decays exponentially along the z direction. Indeed, $\gamma_1$ tends towards $i|\mathbf{k}_{\parallel}|$ for high spatial frequencies.

## 4. A link between the intensity and the sample structure :
## The concept of the transfer function

### 4.1 A NEAR-FIELD HOLOGRAM

We have introduced in section 3 a formalism dealing with the scattering by topography and inhomogeneities. We have shown that the problem can be reduced to the study of scattering by an equivalent rough surface. An expression of the scattered field has been derived. In section 2, we have shown that a detector yields a signal which is linearly related to the intensity (square modulus of the electric field). The intensity at a point $\mathbf{r}$ to first order can be written:

$$I(\mathbf{r}) = \left|\mathbf{E}^{(0)}(\mathbf{r}) + \mathbf{E}^{(1)}(\mathbf{r})\right|^2 \approx \left|\mathbf{E}^{(0)}(\mathbf{r})\right|^2 + 2\,\text{Re}\left[\mathbf{E}^{(0)}(\mathbf{r})\cdot\mathbf{E}^{(1)^*}(\mathbf{r})\right] \tag{12}$$

The first term is the square modulus of the illuminating field. In the case of a plane wave, either homogeneous or inhomogeneous, this term is invariant along the sample surface and depends only on z. Thus this term only contributes to the continuous background of the signal. The second term is an interference term between the illuminating field and the scattered field. This simple remark has two very important consequences.

Assume for instance that the sample is a perfectly plane surface, except for an extremely localized bump that can be viewed as a point-like scatterer. The scattered field is essentially a spherical wave. It turns out that the interference pattern between the spherical wave and the illuminating plane wave produces an extended interference pattern. *A localized structure produces an extended image.* Under plane wave illumination, the relation between the sample and the near-field intensity is *non-local.*

The second important consequence of the interference structure of the second term of Eq.(12) is that the phase of the scattered field has been encoded in an intensity modulation. In other words, the intensity in the near field, produced by the interference between the scattered field and the reference beam, is nothing more than a hologram of the sample. Besides, since the interference is detected in near field, the hologram conveys all the spatial frequencies. This property is essential in view of the solution of the inverse problem. Before dealing with this topic, we shall introduce the concept of the transfer function.

### 4.2 TRANSFER FUNCTION AND IMPULSE RESPONSE

The simplification in Eq.(12) allows us to establish a linear relationship between the surface profile and the modulated part of the intensity. Using Eqs (10-12), the Fourier transform of the intensity can be cast in the form

$$I^{(1)}(\mathbf{k}_{\|}, \mathbf{k}_{\|}^{inc}, z_0) = H_c(\mathbf{k}_{\|}, \mathbf{k}_{\|}^{inc}, z_0) S(\mathbf{k}_{\|}) \tag{13}$$

In this equation, we have considered the intensity along a line at a constant height $z_0$ when the illumination is a plane wave characterized by the wave vector $\mathbf{k}^{inc}$. The tip sample distance $z_0$ controls the spatial frequency cut-off of the transfer function $H_c$ given in appendix A. Eq.(12) can be written in direct space

$$\begin{aligned} I^{(1)}(\mathbf{r}_{\|}, \mathbf{k}_{\|}^{inc}, z_0) &= \int I^{(1)}(\mathbf{k}_{\|}, \mathbf{k}_{\|}^{inc}, z_0) \exp[-i\mathbf{k}_{\|}\mathbf{r}_{\|}] d\mathbf{k}_{\|} \\ &= \int H_c(\mathbf{r}_{\|} - \mathbf{r'}_{\|}, \mathbf{k}_{\|}^{inc}, z_0) S(\mathbf{r'}_{\|}) d\mathbf{r'}_{\|} \end{aligned} \tag{14}$$

The Fourier transform of the transfer function is denoted $H_c(\mathbf{r}_{\|}, \mathbf{k}_{\|}^{inc}, z_0)$ and will be called the impulse response (IR). This quantity describes the non-local link between the sample structure and the intensity. An image reproducing the surface profile is obtained if the IR is a Dirac distribution. A detailed study of the transfer function and the IR can give an understanding of the distorsions produced by the scattering.

12

Consider for instance the case of a plane wave illumination. The IR is shown in Fig.7 a-b for s and p-polarization for an angle of incidence of 0° and 45° respectively. The lower medium has an index of 1.5. The observation distance is $z_0=\lambda/20$. Note that the IR is narrower for p-polarization, indicating that in principle the resolution is better. Another interesting feature is the dissymmetry of the IR for an angle of incidence other than zero.

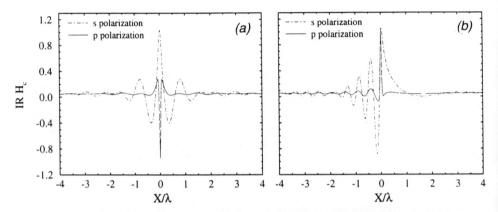

*Figure 7*. Impulse Response for plane wave illumination. (a) $\theta i=0°$ (b) $\theta i=45°$

In Fig.8 a-b, we show the effect of the distance $z_0$ for s and p-polarization, respectively, for normal incidence. It appears clearly that when the distance $z_0$ increases, the IR becomes broader and therefore the resolution decreases. A remarkable feature is that the s and p IR are different at $z_0=0.05\lambda$ and almost identical at $z_0=\lambda$. This behavior is somewhat less surprising when comparing the transfer functions in k-space (not shown here). Indeed, the differences between the two polarizations are more pronounced for high spatial frequencies than for low spatial frequencies. Since propagation acts as a low pass filter, the differences tend to decrease. To conclude, we would like to point out that the concepts of transfer function and IR are a convenient tool for describing and analysing the imaging process.

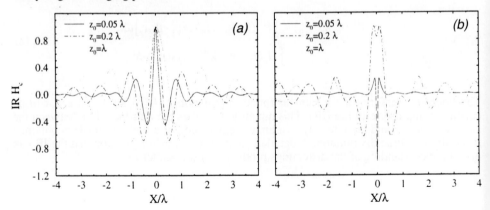

*Figure 8*. IR for increasing $z_0$. (a) s-polarization (b) p-polarization.

## 4.3 ANALYSIS OF THE DISTORSIONS INTRODUCED BY PLANE WAVE ILLUMINATION

Since the first images were produced, it has become clear that the images of a surface profile are very different from the profile itself. For instance, de Fornel *et al.* [34] have made images of a step on a quartz substrate and have shown that the images exhibited large oscillations and were highly dependent on the illumination conditions. This behavior has been analysed in detail in ref. [11]. In this section, we will show that the IR is a convenient tool for understanding these effects. The intensity is a convolution between the surface profile and the IR. Thus, any distorsion observed in the image should appear in the IR. See for instance the intensity pattern shown in Fig. 5b. It corresponds to the convolution of the IR for p-polarization shown in Fig. 7a with the shape of the surface profile. In the case of a classical PSTM, the sample is illuminated by a single plane wave under total internal reflection. The resulting IR is higly asymmetric as seen in Fig.7 b. By contrast, if the incidence is normal (see Fig.7 a), the dissymmetry disappears but the IR is still far from a Dirac distribution. The physical reason for this dissymmetry is the shift in spatial frequency discussed in section 3.4. As can be expected, the shift is zero for normal incidence.

It has been reported recently by Bainier *et al.* [35] that by illuminating with two plane waves with opposite $k_{\parallel}^{inc}$, the distorsion is reduced. The main feature of this system is that the symmetry of the illumination is restored. Yet, there is still a well-defined plane of incidence. Moreover, when using two coherent plane waves, interferences are produced and parts of the sample are not illuminated. Therefore, the image is still highly distorted.

Illumination with a coherent converging beam has not been used in the context of PSTM. Yet, it has been used when the detection is performed by bringing a scattering tip close to the surface [36, 37, 38, 39]. For this case, the link between the intensity and the surface profile cannot be cast in the form of a convolution, the fundamental reason being that the illumination does not have translational invariance. The same remark holds for the illumination recently used by Bainier *et al.* [35]. Except for the case of plane wave illumination, the use of coherent light, such as two interfering plane waves or a converging coherent beam, produces an intensity which is not translationally invariant. Yet, a more general kernel relating the surface profile to the intensity could be derived.

## 4.4 ADVANTAGES OF SPATIALLY PARTIALLY INCOHERENT ILLUMINATION

The incident beam can be characterized by its temporal coherence and its spatial coherence. The usual illuminating beam in PSTM is a laser beam whose spatial (transverse) coherence length is the size of the beam and whose temporal (longitudinal) coherence length is roughly given by $c/\Delta\nu$ where $\Delta\nu$ is the spectral width of the source. Chabrier *et al.* [40] have published results obtained with a rotationally invariant incoherent illumination and using all the angles of incidence larger than the critical angle. They have found that the image was closer to the actual structure of the sample than with the standard PSTM. We have computed in Fig.9 a the corresponding IR for incoherent monochromatic light. In agreement with the reported results, we have found a symmetric more localized IR than for normal incidence (Fig.7a). This result draws the attention to the effects of coherence.

A question of interest in optical imaging is whether coherence is an advantage or a drawback. For most applications of near-field optics, the spectral information is essential. Therefore, an interesting system should use a monochromatic source, which has a perfect temporal coherence. We shall show that spatial coherence can be reduced without loosing useful information.

Recently, Garcia and Nieto-Vesperinas [7] have proposed adding different images obtained with different angles of incidence from -90° to +90°, for a 2D problem. They have shown that this technique reduces spurious oscillations and thus gives a final image which reproduces the surface profile. It turns out that their suggestion is equivalent to taking a single image using a spatially incoherent monochromatic illumination of the sample [8] obtained by superposing a set of incoherent plane waves. A detailed proof of this equivalence is given in Appendix B. For the more general 3D case, the illumination should be rotationaly invariant as in the experimental set-up of ref. [40].

The IR is also very useful in showing the consequences of modifying the illumination. The incoherent IR, $H_{in}$, for the incoherent illumination is obtained from $H_c$ (Eq.13) by integrating over all possible angles of incidence (see Appendix B). Fig.9 b displays the IR obtained for a 2D problem. The lower medium has an index n=1.5 and the observation distance is z0=0.05 $\lambda$. Compared with Fig.7, it is clear that the width has been significantly reduced. Note also that the IR is narrower for p than for s-polarization (the FWHM is $\lambda/10$). An important consequence of this property is that the relationship between the surface and the intensity tends to be local for such an illumination. Note that the incoherent IR (Fig.9 b) integrated over all possible angles of incidence exhibits less oscillations than the incoherent IR (Fig.9a) integrated over the angles of incidence beyond the critical angle. Therefore, a better resolution is expected with an illumination over all possible incidences.

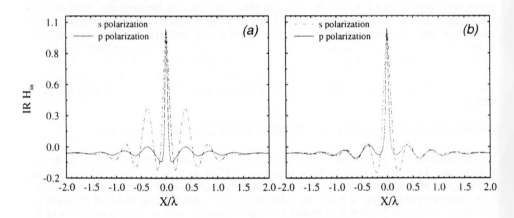

*Figure 9.* Impulse response functions for (a) partially spatially coherent illumination in total internal reflection (b) partially spatially coherent illumination

The width of the IR (or of the transfer function in k-space) describes the resolution capabilities of the technique. It is instructive to look at the cut-off frequency of the transfer function. Fig.10 a-b shows the transfer function for s and p polarization, respectively, for two different indices (n=1.5 and n=4) of the lower medium, through which the sample is illuminated. The observation distance is $z_0=0.05$ $\lambda$. The cut-off frequency is larger for an index of 4. The reason for this difference lies in the maximum spatial frequency available in the spectrum of the illuminating light, $n\omega/c$. It can be seen upon inspection of Eq.(11) that the maximum spatial frequency of the surface which is not exponentially attenuated is given by $|\mathbf{k}_{\parallel}-\mathbf{k}_{\parallel}^{inc}|<\omega/c$. Since the maximum value of $|\mathbf{k}_{\parallel}^{inc}|$ is $n\omega/c$, the maximum spatial frequency $|\mathbf{k}_{\parallel}|$ is expected to be $(1+n)\omega/c$. The value of $(1+n)\omega/c$ is indicated by vertical lines in Fig.10. Note that this idea is the essence of immersion microscopy.

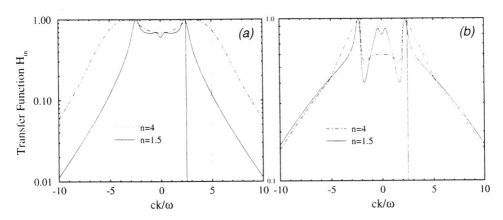

*Figure 10.* Incoherent transfer function $H_{in}$ for two values of the index of the substrate (a) s polarization (b) p polarization.

To put forward the advantage of imaging with incoherent isotropic illumination, we have considered a small structure composed of two ridges separated by $\lambda/10$, as depicted in Fig.11 a. Their height is $\lambda/40$ and their width is $\lambda/4$. The intensity at constant height $z_0=0.0625\lambda$ obtained under the standard PSTM configuration with an illuminating plane wave at 45° is shown in Fig.11 c. The corresponding result obtained by incoherently superimposing the intensities produced by a set of 43 plane waves equally spaced between -84° and +84° is shown in Fig.11 d. It is clear that the incoherent illumination provides a means to retrieve the surface profile with a super-resolution capability.

4.5 ILLUMINATION CONDITIONS

Let us briefly conclude the analysis of the illumination conditions. A simple convolution relation exists between the surface profile and the intensity for incoherent illumination or plane wave illumination. In this case, one can define an IR that conveys all the information about the image formation. Note that the spatial incoherence prevents interference between the illuminating plane waves and therefore ensures the translational invariance of the illumination. This translational invariance in turn guarantees the existence of an IR. Of course, for particular problems, the classical plane wave illumination may be better suited, such as for the excitation of a surface plasmon polariton along a given direction [41, 42, 43]. Yet, examination of the corresponding IR (Fig. 7 b) clearly demonstrates that this sort of illumination is not appropriate for general image formation.

16

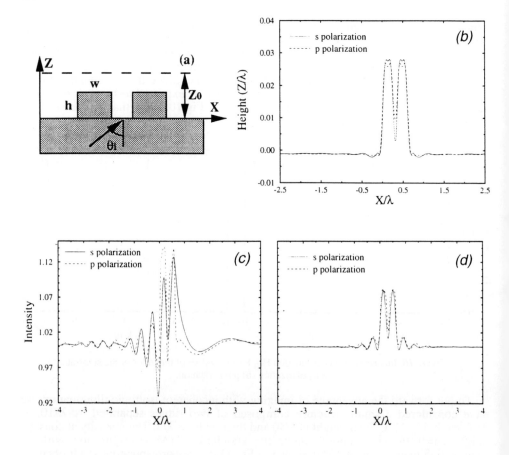

*Figure 11.* Spatially incoherent imaging. (a) geometry (b) surface profile after deconvolution (c) near-field intensity at constant height with coherent illumination (d) same as c) with spatially incoherent illumination .

The image distorsions are directly related to the symmetry of the illumination. The IR has to be rotationally invariant if we wish to suppress the distorsions as much as possible between the image and the equivalent surface profile. Another issue is the resolution which is described by the width of the IR. Ideally, the IR should be a Dirac distribution which is equivalent to having a transfer function which is constant in k-space. This would ensure a local relation between the equivalent surface profile and the intensity. The above analysis shows that the transfer function starts decaying significantly beyond $(1+n)\omega/c$ for an isotropic illumination. Therefore, the best monochromatic illumination is *an incoherent transmission illumination through a high index material with rotational invariance along the direction normal to the sample.*

4.6 SOLUTION OF THE INVERSE PROBLEM

We have discussed the implications of spatially incoherent illumination and shown that it leads to an almost local relationship between the surface profile and intensity. Thus, the image approaches the actual structure. Yet, the value of the height of the equivalent surface profile is not obtained with this procedure. Moreover, the resolution is limited by the width of the IR. The inversion procedure introduced in Ref.[6] can still be applied to quantitatively reconstruct the equivalent surface profile. The idea is to deconvolve the intensity, taking advantage of the fact that the IR is analytically known. In practice, the Fourier transform of the intensity is computed and divided by the transfer function to obtain the surface profile spectrum. An inverse Fourier transform yields quantitatively the equivalent surface profile. To illustrate the further improvement provided by this procedure, we show in Fig.11 b the surface profile obtained after reconstruction using Fig.11 d as input data. The value of the height is quantitatively retrieved and the angles of the profile are better reconstructed. Note also that the reconstruction procedure can be applied to data generated either with a coherent plane wave (Ref.[6]) or with an incoherent beam.

Reconstruction is possible for incoherent light even though the basic phenomenon is the interference between the illuminating beam and the scattered field. This is equivalent to a white-light hologram, which is possible because all the optical paths involved are very short.

## 5. Speckle structure in the near field

A number of authors have pointed out that the quality of the images depend drastically on the coherence of the illumination [31, 44, 45]. In this section, we focus our attention on the features of the images that are produced by the residual random roughness of the surface. Basically, a random roughness scatters the incident light and produces a speckle pattern. In this section, we will introduce the tools necessary for describing both a random rough surface and a speckle pattern. Then, we will derive a simple relationship between the speckle pattern and statistical properties of the random rough surface.

5.1 STATISTICAL DESCRIPTION OF A RANDOM ROUGH SURFACE

Given a surface profile defined by a function $z=S(\mathbf{r}_{\parallel})$. We assume that $S(\mathbf{r}_{\parallel})$ is a random variable with zero mean value and a Gaussian height probability density function. The ensemble average is denoted by brackets and the transverse structure of the surface is described by the correlation function $C(|\mathbf{r'}_{\parallel}-\mathbf{r}_{\parallel}|)$ :

$$<S(\mathbf{r}_{\parallel})> = 0 \tag{15}$$

$$<S(\mathbf{r}_{\parallel})S(\mathbf{r'}_{\parallel})> = \delta^2 C(|\mathbf{r'}_{\parallel}-\mathbf{r}_{\parallel}|) \tag{16}$$

$\delta$ denotes the rms height. We assume an isotropic stationary random process as seen in the form of the correlation function which depends only on the distance $|\mathbf{r}_{\parallel}-\mathbf{r}_{\parallel}'|$.

The Fourier transform of the correlation function $g(\mathbf{k}_{||})$ is defined by

$$C(|\mathbf{u}_{||}|) = \int d\mathbf{k}_{||} g(\mathbf{k}_{||}) \exp[i\mathbf{k}_{||}\mathbf{u}_{||}] \tag{17}$$

The stationarity introduces some difficulties to define the Fourier transform of the surface profile. To overcome this problem, we introduce a function $S_A(\mathbf{r}_{||})$ which is equal to the surface profile within the area A of the x-y plane and zero outside. The Fourier transform of the truncated surface is given by

$$S_A(\mathbf{r}_{||}) = \int \frac{d\mathbf{k}_{||}}{4\pi^2} S_A(\mathbf{k}_{||}) \exp[-i\mathbf{k}_{||}\mathbf{r}_{||}] \tag{18}$$

The Wiener Khinchin theorem relates the Fourier transform of the correlation function $g(\mathbf{k}_{||})$ to the power spectral density of the surface :

$$\lim_{A \to \infty} \frac{< S_A(\mathbf{k}_{||})S_A(-\mathbf{k}_{||}) >}{A} = 4\pi^2 \delta^2 g(\mathbf{k}_{||}) \tag{19}$$

## 5.2 STATISTICAL DESCRIPTION OF THE NEAR FIELD

The intensity in a plane at a constant height $z=z_0$ is a random variable, characterized by its probability density function. This function can be deduced from the height probability density function of the surface and Eq.(14). The transverse structure of the intensity in the plane $z=z_0$ is described by the correlation function $<I(\mathbf{r}_{||})I(\mathbf{r}_{||}+\mathbf{u})>$. Using the first-order expansion of the intensity introduced in section 4, we obtain :

$$<I(\mathbf{r}_{||})I(\mathbf{r}_{||}+\mathbf{u})> = <I^{(0)}I^{(0)}> + <I^{(1)}(\mathbf{r}_{||})I^{(1)}(\mathbf{r}_{||}+\mathbf{u})> \tag{20}$$

The first term is a constant for a plane wave illumination. The first order terms vanish due to Eq.(15) which implies $<I^{(1)}(\mathbf{r}_{||}+\mathbf{u})>=0$. Note that we have neglected the term $<I^{(0)}(\mathbf{r}_{||})I^{(2)}(\mathbf{r}_{||}+\mathbf{u})>$ because $I^{(0)}$ is deterministic so that it can be factorized as $<I^{(0)}(\mathbf{r}_{||})><I^{(2)}(\mathbf{r}_{||})>$ and does not depend on $\mathbf{u}$. A useful quantity to characterize the speckle pattern is the speckle contrast, defined as the rms fluctuation of the intensity normalized by the mean intensity. To lowest order, it is given by

$$C_I = \frac{\sqrt{\langle I^2 \rangle - \langle I \rangle^2}}{\langle I \rangle} = \frac{\sqrt{\langle I^{(1)2} \rangle}}{I^{(0)}} \tag{21}$$

For most imaging applications, a low speckle contrast is desirable. In the following, we will use the formalism of section 3 to analyse the link between the near-field speckle and the statistical properties of the surface.

## 5.3 ANALYSIS OF THE LINK BETWEEN THE SPECKLE PATTERN AND THE STATISTICAL PROPERTIES OF THE SURFACE.

The starting point of the analysis is Eqs.(10-12). Since $I^{(1)}$ is linearly related to the surface, the correlation function $<I^{(1)}(r_{||})I^{(1)}(r_{||}+u)>$ is quadratic in S. The detailed expression can be cast in the form :

$$\left\langle I^{(1)}(r_{||})I^{(1)}(r_{||}+u)\right\rangle = \int \delta^2 g(k_{||}) H_I(k_{||}, k_{||}^{inc}, z_0) \exp[ik_{||}.u]dk_{||} \tag{22}$$

where $H_I(k_{||}, k_{||}^{inc}, z_0)$ is a transfer function that depends on the polarization, the height of the observation point, and the wave vector of the incident field. The above expression shows that $<I^{(1)}(r_{||})I^{(1)}(r_{||}+u)>$ is proportional to $\delta^2$. Therefore the speckle contrast is proportional to the rms height $\delta$.

The transfer function should not be confused with the transfer function introduced in section 4. The latter was relating the intensity to the surface spectrum whereas $H_I$ connects the intensity correlation function with the power spectral density of the surface profile. Its detailed expression is given in appendix A. Again, this transfer function is completely known provided that the illumination conditions are known. Therefore, a measurement of the correlation of the near-field intensity allows to compute the power density spectrum of the surface itself. The main advantage as compared with far-field scattering techniques is that the knowledge of $g(k_{||})$ is no longer band limited. This property may find interesting applications for surface characterization when correlation lengths are in the micronic range.

### 5.4 EFFECT OF THE COHERENCE ON THE SPECKLE PATTERN

The aim of this section is to present the effect of a partially coherent illumination on the speckle pattern. We will simulate the intensity in the near field for 3D geometries using the perturbative formalism of section 3. The generation of a random rough surface has been done using the algorithm described in ref. [46]. The surface profile is shown in Fig.12 a. The object is a parallelepiped of width 1200 nm and height 15 nm. A residual roughness (rms $\delta=1$ nm and correlation length a=120 nm) has been added on the surface. The material has a dielectric constant $\varepsilon_2=2.25$, the upper medium is a vaccum. The intensity along a plane at a constant height $z_0=30$ nm above the surface produced by a monochromatic plane wave ($\lambda=600$ nm) illuminating the surface in total internal reflection ($\theta_i=45°$) is shown in Fig.12 b. The plane of incidence is y-z. The image exhibits a speckle pattern which is added on the image of the object. Note that the image loses its symmetry, due to the fact that the illumination, is not rotationally invariant around the z-axis.

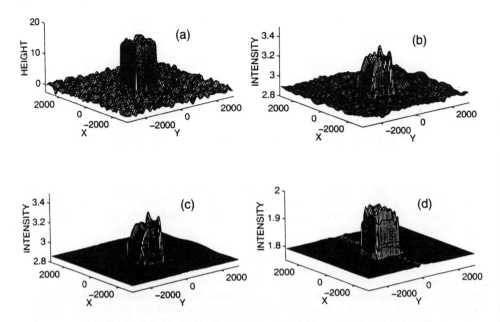

*Figure 12*. (a) Surface profile with a residual roughness (b) Near-field intensity above the surface with a coherent illumination (c) same as b) with a reduced temporal coherence (d) same as b) with a reduced spatial coherence

A sandard way to decrease the speckle contrast in far-field imaging is to reduce the coherence of the illumination. Note that the transfer function $H_I$ introduced above depends both on the incident wave vector and on the incident frequency. Therefore, it conveys all the information on the spatial and temporal coherence of the illumination. It can be integrated over $k_{inc}$ to yield a partially spatially coherent function and over $\omega$ for a partially temporal coherent function.

The intensity under illumination with a white light source ($400<\lambda<800$ nm) and a well defined incidence direction ($\theta_i=45°$) is shown in Fig.12 c. Note that in this case, we have reduced the temporal coherence of the illuminating wave. To reduce the spatial coherence, we have used a monochromatic beam composed of several uncorrelated plane waves with different angles of incidence but with the same plane of incidence y-z. The resulting intensity is shown in Fig.12 d.

As seen in Fig.12 c-d, when using a partially coherent source, the speckle contrast is significantly attenuated. This result was reported experimentally by several authors [11, 44, 45]. It is interesting to note that the image under spatially incoherent illumination (Fig.12 d) is more symmetric than in the case of an illumination with one plane wave, as discussed previously, although the illumination was not rotationaly invariant (the plane of incidence was the same for each plane wave of the beam). With a perfectly symmetric illumination, the restoration of the object shape would have been better.

## 6. Conclusion

An analysis of the image formation in detection mode near-field optical microscopy has been presented. The analysis has been restricted to cases for which the tip is a passive probe. This issue has been discussed and it has been shown that this assumption encompasses a wide range of situations. Under this assumption, the coupling between the tip and the field can be accounted for by a transfer function in a non-perturbative manner. Throughout the text we have assumed that the relevant quantity detected by the tip is the intensity of the near field. Although this hypothesis is consistent with numerical simulations and experiments, we believe that this issue is not completely resolved, particularly for 3D systems.

Relying on the passive probe assumption, we have focussed our attention on the link between the surface profile and the intensity. This problem can be analysed in the framework of perturbation theory. This approach yields a physical insight into the image formation process. We have shown that the near-field intensity is related to both the surface profile of the sample and to the dielectric contrast through a coupled term. This fact implies that different samples with different surface profiles and dielectric contrast may produce the same image. Therefore, one may define an equivalent surface profile that accounts for both topography and dielectric contrast. This property, predicted on the basis of perturbation theory, has been confirmed by rigorous numerical simulations.

Another significant result of perturbation theory is that the intensity in the near field results from interferences between the illuminating beam and the scattered field. Therefore, the intensity pattern is a hologram of the sample. Since it is a near-field hologram involving evanescent waves, it yields information over a wide range of spatial frequencies. This result indicates that the link between the surface profile and the sample is non-local and that both the phase and the amplitude of the scattered field are encoded in the intensity pattern allowing a reconstruction of the equivalent surface profile.

Within the perturbative framework, it is possible to gain a simple and complete picture of the image formation process by introducing the concepts of the transfer function in k-space and the Impulse Response in direct space. The basic result is that the intensity is a convolution product of the equivalent surface profile and the IR. This is a consequence of the linear relationship between the equivalent surface profile of the sample and the near-field intensity, due to the perturbative approach. The IR provides a description of the bias produced in the images by a plane wave illumination. It shows the importance of the symmetry of the illumination in order to obtain a symmetric IR. The effect of polarization, sample-tip distance and coherence of the illumination can be discussed in simple terms with this concept. We have shown that, for a general imaging purpose, plane wave illumination is not well suited. Instead, a monochromatic, partially incoherent, symmetric illumination produces images which are close to the actual structure of the sample, due to the fact that the relationship between the sample and the intensity becomes local.

The IR, which is analytically known, can be used to perform a deconvolution of the signal in order to recover the equivalent surface profile. This reconstruction procedure improves the resolution of the image. Besides, it yields quantitative information on the height of the equivalent surface profile.

A feature that has been observed in near-field images is a kind of near-field speckle pattern produced by the residual random roughness of the sample. Since the rms height is usually less than a few nanometers, perturbative theory can be used to describe the scattered field. It turns out that the power spectral density of the surface profile and the near-field intensity are related by a filter. We have shown that : 1) the near-field speckle contrast is proportional to the rms height, 2) as for the far-field speckle, the near-field speckle contrast decreases with decreasing the coherence of the illumination.

## 7. Appendix A

### 7.1 COHERENT TRANSFER FUNCTION $H_C$

Using Eqs (10-12) it can be shown that the first order intensity can be cast in the form :

$$I^{(1)}(\mathbf{r}_\parallel, \mathbf{k}_\parallel^{inc}, z_0) = \int H_c(\mathbf{k}_\parallel, \mathbf{k}_\parallel^{inc}, z_0) S(\mathbf{k}_\parallel) \exp\left[-i\mathbf{k}_\parallel \mathbf{r}_\parallel\right] d\mathbf{k}_\parallel \qquad (A\text{-}1)$$

where $H_c(\mathbf{k}_\parallel, \mathbf{k}_\parallel^{inc}, z_0)$ is given by

$$H_c(\mathbf{k}_\parallel, \mathbf{k}_\parallel^{inc}, z_0) = f(\mathbf{k}_\parallel^{inc} - \mathbf{k}_\parallel, \mathbf{k}_\parallel^{inc}, z_0) + f^*(\mathbf{k}_\parallel^{inc} + \mathbf{k}_\parallel, \mathbf{k}_\parallel^{inc}, z_0) \qquad (A\text{-}2)$$

and

$$f(\mathbf{k}_\parallel, \mathbf{k}_\parallel^{inc}, z_0) = \left[t^*\left(\mathbf{k}_\parallel^i\right)\right] \mathbf{e}_{inc}^* \cdot L\left(\mathbf{k}_\parallel, \mathbf{k}_\parallel^i\right) \mathbf{e}_{inc} \exp\left[i\gamma_1(\mathbf{k}_\parallel)z_0 - i\gamma_1 *\left(\mathbf{k}_\parallel^i\right)z_0\right] \qquad (A\text{-}3)$$

where the asterisk denotes a complex conjugate. The operator $L(\mathbf{k}_\parallel, \mathbf{k}^i_\parallel)$ is explicitly given in Ref. [11, 31] and $[t(\mathbf{k}^i_\parallel)]$ is a matrix of transmission Fresnel factors.

### 7.2 FILTER CONNECTING THE POWER SPECTRAL DENSITY OF THE INTENSITY TO THE POWER SPECTRAL DENSITY OF THE SURFACE

Using Eqs (10-12), (19) and (20) it can be shown that the filter is given by

$$H_I\left(\mathbf{k}_\parallel, \mathbf{k}_\parallel^{inc}, z_0\right) = F\left(\mathbf{k}_\parallel, \mathbf{k}_\parallel^{inc}, z_0\right) + F^*\left(-\mathbf{k}_\parallel, \mathbf{k}_\parallel^{inc}, z_0\right) \qquad (A\text{-}4)$$

where the function F is

$$F(\mathbf{k}_{\parallel}, \mathbf{k}_{\parallel}^{inc}, z_0) = \left[ \mathbf{e}_t^{(0)*} \cdot L(\mathbf{k}_{\parallel}^{inc} - \mathbf{k}_{\parallel}, \mathbf{k}_{\parallel}^{inc}) \mathbf{e}_{inc} \right] \exp\left[ i\gamma_1 (\mathbf{k}_{\parallel}^{inc} - \mathbf{k}_{\parallel}) z_0 \right]$$

$$\left\{ \begin{array}{l} \mathbf{e}^{(0)} \cdot L^*(\mathbf{k}_{\parallel}^{inc} - \mathbf{k}_{\parallel}, \mathbf{k}_{\parallel}^{inc}) \mathbf{e}_{inc}^* \exp\left\{ -2\,\mathrm{Im}\left[ \gamma_1(\mathbf{k}_{\parallel}^{inc}) \right] z_0 - i\gamma_1^*(\mathbf{k}_{\parallel}^{inc} - \mathbf{k}_{\parallel}) z_0 \right\} + \\ \mathbf{e}^{(0)*} \cdot L(\mathbf{k}_{\parallel}^{inc} + \mathbf{k}_{\parallel}, \mathbf{k}_{\parallel}^{inc}) \mathbf{e}_{inc} \exp\left[ -2i\gamma_1^*(\mathbf{k}_{\parallel}^{inc}) z_0 + i\gamma_1(\mathbf{k}_{\parallel}^{inc} + \mathbf{k}_{\parallel}) z_0 \right] \end{array} \right\} \qquad \text{(A-5)}$$

## 8. Appendix B

Let the incident field be a sum of monochromatic plane waves that are incoherent (a temporal dependence $\exp(-i\omega t)$ is assumed for all fields).

$$\mathbf{E}_{inc}(\mathbf{r}_{\parallel}, z_0) = \int \mathbf{e}_{inc}(\mathbf{k}_{\parallel}^i) \exp\left[ i\mathbf{k}_{\parallel}^i \mathbf{r}_{\parallel} + i\gamma(\mathbf{k}_{\parallel}^i) z_0 \right] d\mathbf{k}_{\parallel}^i \qquad \text{(B-1)}$$

where $\mathbf{e}_{inc}(\mathbf{k}^i{}_{\parallel})$ is a random variable with an angular delta correlation function :

$$\left\langle \mathbf{e}_{inc}(\mathbf{k}_{\parallel}^i) \cdot \mathbf{e}^*{}_{inc}(\mathbf{k}_{\parallel}^{i\prime}) \right\rangle = |\mathbf{e}_{inc}|^2 \delta(\mathbf{k}_{\parallel}^i - \mathbf{k}_{\parallel}^{i\prime}) \qquad \text{(B-2)}$$

The superscript * denotes the conjugate of a complex number. We assume in the following that $|\mathbf{e}_{inc}|$ depends on $|\mathbf{k}^i{}_{\parallel}|$ which amounts to assume that the illumination has a rotational invariance around the z-axis. To first order in perturbation the intensity is given by Eq. (12). In this equation the zero order field is given by

$$\mathbf{E}^{(0)}(\mathbf{r}_{\parallel}, z_0) = \int \left[ t(\mathbf{k}_{\parallel}^i) \right] \mathbf{e}_{inc}(\mathbf{k}_{\parallel}^i) \exp\left[ i\mathbf{k}_{\parallel}^i \mathbf{r}_{\parallel} + i\gamma(\mathbf{k}_{\parallel}^i) z_0 \right] d\mathbf{k}_{\parallel}^i \qquad \text{(B-3)}$$

and the first order field is given by

$$\mathbf{E}^{(1)}(\mathbf{r}_{\parallel}, z_0) = \int d\mathbf{k}_{\parallel}^i \int L(\mathbf{k}_{\parallel}, \mathbf{k}_{\parallel}^i) \mathbf{e}_{inc}(\mathbf{k}_{\parallel}^i) S(\mathbf{k}_{\parallel}^i - \mathbf{k}_{\parallel}) \exp\left[ i\mathbf{k}_{\parallel} \mathbf{r}_{\parallel} + i\gamma(\mathbf{k}_{\parallel}) z_0 \right] d\mathbf{k}_{\parallel} \qquad \text{(B-4)}$$

Using Eqs (12) and (B-1)-(B-4) one can cast the intensity in the form :

$$I(\mathbf{r}_{\parallel}, z_0) = \left| \mathbf{E}^{(0)}(\mathbf{r}_{\parallel}, z_0) \right|^2 + \int H_{in}(\mathbf{k}_{\parallel}, z_0) S(\mathbf{k}_{\parallel}) \exp\left[ -i\mathbf{k}_{\parallel} \mathbf{r}_{\parallel} \right] d\mathbf{k}_{\parallel} \qquad \text{(B-5)}$$

The first term in the expression (B-5) of the intensity is a function of $z_0$ only. The filter appearing in the second term is given by

$$H_{in}(\mathbf{k}_{\parallel}, z_0) = F(\mathbf{k}_{\parallel}, z_0) + F^*(-\mathbf{k}_{\parallel}, z_0) \qquad \text{(B-6)}$$

where

$$F(\mathbf{k}_{\shortparallel}, z_0) =$$

$$\int L(\mathbf{k}_{\shortparallel}^i - \mathbf{k}_{\shortparallel}, \mathbf{k}_{\shortparallel}^i) e_{inc}(\mathbf{k}_{\shortparallel}^i) \cdot [t^*(\mathbf{k}_{\shortparallel}^i)] e_{inc}^*(\mathbf{k}_{\shortparallel}^i) \exp[i\gamma(\mathbf{k}_{\shortparallel}^i - \mathbf{k}_{\shortparallel}) z_0 - i\gamma * (\mathbf{k}_{\shortparallel}^i) z_0] d\mathbf{k}_{\shortparallel}^i \tag{B-7}$$

Equation (B-5) possesses the same structure as Eq.(14). It demonstrates that the modulated part of the intensity at constant height reproduces the surface profile provided that the filter does not modify drastically the spectrum. This filter is the transfer function extensively discussed in section 4. For incoherent symetric illumination, it yields a narrow symetric impulse response. Note that the filter obtained through this calculation is exactly the integral over the incident wave vector of the coherent filter derived in appendix A. This result is trivial if one assumes that the intensities produced by each plane wave can be summed. In this appendix, we have proved that this assumption is correct if the field has a delta angular correlation as expressed by Eq.(B-2).

### 9. References

1. Pohl, D.W. (1992) Nano-optics and Scanning Near-Field Optical Microscopy, in R. Wiesendanger and H.J.Güntherodt (eds.), *Scanning Tunneling Microscopy II*, Springer-Verlag, Berlin, pp. 233-271.
2. Heinzelmann, H. and Pohl, D.W. (1994) Scanning near-field microscopy, *Appl. Phys. A* **59**, 89-101.
3. Courjon, D. and Bainier, C. (1994) Near-field microscopy and near-field optics, *Rep. Prog. Phys.* **57**, 989-1028.
4. Betzig, E., Lewis, A., Harootunian, A., Isaacson, M. and Kratschmer, E. (1986) Near-field scanning optical microscopy. Development and biophysical applications, *Biophys. J.* **49**, 269-279.
5. Garcia, N. and Nieto-Vesperinas, M. (1993) Near-Field optics inverse-scattering reconstruction of reflective surfaces, *Opt. Lett.* **18**, 2090-2092.
6. Greffet, J.-J., Sentenac, A. and Carminati, R. (1995) Surface profile reconstruction using near field data, *Opt. Commun.* **116**, 20-24.
7. Garcia, N. and Nieto-Vesperinas, M. (1995) Direct solution to the inverse scattering problem for surfaces from near-field intensities without phase retrieval, *Opt. Lett.* **20**, 949-951.
8. Carminati, R., Greffet, J.-J., Garcia, N. and Nieto-Vesperinas, M. (1996) Direct reconstruction of surfaces from near-field intensity under spatially incoherent illumination, *Submitted for publication* .
9. Totzeck, M. and Krumbügel, M.A. (1995) Test of various diffraction theories in the near field of phase objects, *Ultramicroscopy* **57**, 160-164.
10. Totzeck, M. and Krumbügel, M.A. (1994) Lateral resolution in the near field and far field phase images of π-phase shifting structures, *Opt. Commun.* **112**, 189-200.
11. de Fornel, F., Adam, P.M., Salomon, L., Goudonnet, J.-P., Sentenac, A., Carminati, R. and Greffet, J.-J. (1996) Analysis of the image formation with a Photon Scanning Tunneling Microscope, *J. Opt. Soc. Am. A* (in press).
12. Carminati, R. and Greffet, J.J. (1995) Two dimensional numerical simulation of the Photon Scanning Tunneling Microscope. Concept of transfer function, *Opt. Commun.* **116**, 316-321.

13. Carminati, R. and Greffet, J.J. (1995) Influence of dielectric contrast and topography on the near field scattered by an inhomogeneous surface, *J. Opt. Soc. Am. A* **12**, (in press).
14. Keller, O. (1993) Optical near-field interaction : on the local field inside a quantum tip, in D. W. Pohl and D. Courjon (eds.), Kluwer, Dordrecht, pp. 379-390.
15. Keller, O. (1994) Optical polarizability of small quatum particles : local-field effects in a self-field approach, *J. Opt. Soc. Am. B* **11**, 1480-1489.
16. Keller, O., Xiao, M. and Bozhevolnyi, S. (1994) Optical paramagnetic polarizability of mesoscopic particles: a study of local fields corrections, *Opt. Commun.* **114**, 491-500.
17. Greffet, J.-J. (1989) Scattering of s-polarized electromagnetic waves by a 2 D obstacle near an interface, *Opt. Commun.* **72**, 274-278.
18. Pincemin, F., Sentenac, A. and Greffet, J.-J. (1994) Near field scattered by a dielectric rod below a metal surface, *J. Opt. Soc. Am. A* **11**, 1117-1127.
19. Martin, O.J.F., Dereux, A. and Girard, C. (1994) Iterative scheme for computing exactly the total field propagating in dielectric structures of arbitrary shape, *J. Opt. Soc. Am. A* **11**, 1073-1080.
20. Girard, C. and Bouju, X. (1991) Coupled electromagnetic modes between a corrugated surface and a thin probe tip, *J. Chem. Phys.* **3**, 2056-2064.
21. Girard, C. and Courjon, D. (1990) Model for scanning tunneling optical microsocpy: A microscopic self-consistent approach, *Phys. Rev. B* **42**, 9340-9349.
22. Lakhtakia, A. (1990) Macroscopic theory of the coupled dipole approximation method, *Opt. Commun.* **79**, 1-5.
23. Agarwal, G.S. (1976) Integral equation treatment of scattering from rough surfaces, *Phys.Rev.B* **14**, 846-848.
24. Agarwal, G.S. (1977) Interaction of electromagnetic waves at rough dielectric surfaces, *Phys.Rev. B* **15**, 2371-2383.
25. Maradudin, A.A. and Mills, D.L. (1975) Scattering and absorption of electromagnetic radiation by a semi-infinite medium in the presence of surface roughness, *Phys. Rev. B* **11**, 1392-1415.
26. Kröger, E. and Kretschmann, E. (1970) Scattering of light by slightly rough surfaces or thin films including plsma resonance emission, *Z. Physik* **237**, 1-15.
27. Nieto-Vesperinas, M. (1982) Depolarization of electromagnetic waves scattered from slightly rough random surfaces : a study by means of the extinction theorem, *J. Opt. Soc. Am.* **72**, 539-547.
28. Greffet, J.-J. (1988) Scattering of electromagnetic waves by rough dielectric surfaces, *Phys. Rev. B* **37**, 6436-6441.
29. Barchiesi, D. and Van Labeke, D. (1992) Scanning tunneling optical microscopy (STOM) : Theoretical study of polarization effects with two models of tip, in D.W. Pohl and D. Courjon (eds.), *Near Field Optics,* Kluwer, Dordrecht, pp. 179-188.
30. Van Labeke, D. and Barchiesi, D. (1992) Scanning-tunneling optical microscopy : a theoretical macroscopic approach, *J. Opt. Soc. Am. A* **9**, 732-739.
31. Greffet, J.-J. and Carminati, R. (1996) Relationship between the near field speckle pattern and the statistical properties of a surface, *Ultramicroscopy* (in press).
32. Sentenac, A. and Greffet, J.J. (1995) Study of the features of PSTM images by means of a perturbative approach., *Ultramicroscopy* **57**, 246-250.
33. Van Labeke, D. and Barchiesi, D. (1993) Theoretical problems in scanning near-field optical microscopy, in D. W. Pohl. and D. Courjon (eds.), *Near Field Optics,* Kluwer, Dordrecht, pp. 157-178.

34. de Fornel, F., Bourillot, E., Adam, P., Salomon, L., Chabrier, G. and Goudonnet, J.P. (1993) Recent experimental results with the PSTM : Observation of a step on a quartz surface. Spatial spectroscopy of microwaveguides, in D. W. Pohl and D. Courjon (eds.), *Near Field Optics,* Kluwer, Dordrecht, pp. 59-70.

35. Bainier, C., Courjon, D. and Baida, F. (1995) Evanescent interferometry by scanning optical tunneling detection, *J. Opt. Soc. Am. A* to be published.

36. Gleyzes, P., Boccara, A.C. and Bachelot, R. (1995) Near field optical microscopy using a metallic vibrating tip, *Ultramicroscopy* **57**, 318-322.

37. Bachelot, R., Gleyzes, P. and Boccara, A.C. (1995) Near-Field optical microscope based on local perturbation of a diffraction spot, *Opt. Lett.* **20**, 1924-1926.

38. Zenhausern, F., O'Boyle, M.P. and Wickramasinghe, H.K. (1994) Apertureless near-field microscope, *Appl. Phys. Lett.* **65**, 1623-1625.

39. Kawata, S. and Inouye, Y. (1995) Scanning probe optical microscopy using a metallic probe tip, *Ultramicroscopy* **57**, 313-317.

40. Chabrier, G., de Fornel, F., Bourillot, E., Salomon, L. and Goudonnet, J.-P. (1994) A dark field scanning tunneling microscope under incoherent light illumination, *Opt. Commun.* **107**, 347-352.

41. Bozhevolnyi, S., Smolyaninov, I.I. and Zayats, A.V. (1995) Near-field microscopy of surface-plasmon polaritons: localization and internal interface imaging, *Phys. Rev. B* **51**, 17916-17924.

42. Bozhevolnyi, S., Vohnsen, B., Smolyaninov, I.I. and Zayats, A.V. (1995) Direct observation of surface polariton localization caused by surface roughness, *Opt. Commun.* **117**, 417-423.

43. Dawson, P., Smith, K.W., de Fornel, F. and Goudonnet, J.-P. (1995) Imaging of surface plasmon launch and propagation using a photon scanning tunneling microscope, *Ultramicroscopy* **57**, 287-292.

44. de Fornel, F., Adam, P.M., Salomon, L., Goudonnet, J.P. and Guérin, P. (1994) Effect of the coherence of the sources on the images obtained with a Photon Scanning Tunneling Microscope", *Opt. Lett.* **19**, 14-17.

45. Toledo-Crow, R., Smith, B.W., Rogers, J.K. and Vaez-Iravani, M. (1994) Near Field Optical Microscopy Characterization of IC Metrology, *S.P.I.E.* **2196**, 62-73.

46. Maradudin, A.A., Michel, T., McGurn, A.R. and Mendez, E.R. (1990) Enhanced backscattering of light from a random grating, *Ann. Phys. (N.Y.)* **203**, 255-307.

# LIGHT SCATTERING BY TIPS IN FRONT OF SURFACES

A. MADRAZO[1], M. NIETO-VESPERINAS[1] AND N. GARCÍA[2]

[1] *Instituto de Ciencia de Materiales, CSIC, and Departamento de Física de la Materia Condensada.*
[2] *Física de Sistemas Pequeños, CSIC-UAM.*
*Facultad de Ciencias C-III. Universidad Autónoma.*
*Cantoblanco.*
*Madrid 28049. Spain.*

**Abstract.** An exact solution to the electromagnetic interaction between a local probe and the surface under study is presented by using a non local boundary condition. Two configurations of current experimental interest have been addressed, one is the plasmon field enhancement behaviour in a metallic tip-sample configuration at subnanometer proximity. The other case involves the detection of subwavelength Goos-Hänchen lateral shifts in a PSTM configuration.

## 1. Introduction

Near Field Optics calculations [1]-[4] involve the difficulty of determining in an exact form the multiple interaction between light, the object surface and the detecting tip. Here we present an exact solution of the wave equation for a tip-interface configuration by using a non local boundary condition. The system is two-dimensional, and consists of a cylinder at subwavelenght distance from a corrugated surface. Both media are described by a frecuency dependent complex dielectric constant.

We shall illustrate two cases of current interest. One is that pertaining to the field between a tip and an object surface, either of them being of W or Ag. The other is the detection of subwavelength Goos-Hänchen shifts. In the first case, the plasmon intensity has a huge enhancement which is maximum when the tip-surface distance is about 1 nm, and it is sensitive to

*M. Nieto-Vesperinas and N. García (eds.), Optics at the Nanometer Scale 27–40.*
© *1996 Kluwer Academic Publishers. Printed in the Netherlands.*

variations of 0.1 nm. This can explain recent experimental data on atomic resolution in photon emission by scanning tunneling microscopy operating at constant current mode by R. Berndt et al. [5]. They have shown atomic resolution in photon emission induced by scanning tunnelling microscopy (STM). In the experiment they look at the light emission from a scanning STM tip on a Au surface while a voltage between 2 and 4 volts is applied between the tip and the surface. The fact that light can be emitted when a current passes through two electrodes was first observed by Lambe and Mc-Carthy[6]. This was later confirmed in a STM configuration[7] and a theory was presented[8]-[9]. However, up till now there exist no rigorous calculations that can predict and show that atomic resolution can be obtained with light emission. The problem is that the emitted light has a wavelength of 600 nm in comparison with the 0.8 nm periodicity of the Au (110) surface observed in Ref. [5]. A theory exists for explaining atomic resolution when the STM experiment is run at constant height of the tip[10], i.e., at a varying tunnel current. Since the tunnel gap is smaller at the maximum of the surface height, this implies that the current is larger in a maximum rather than in a minimum of the surface. However, this theory cannot explain the experiments that are performed at constant current mode and give more light emission when the tip is in the minima than when it is in the maxima of the surface[5], this being just the opposite of the theoretical predictions[10]. Therefore, a theory that explains the contrast observed experimentally is necessary, namely, that there is contrast at constant current and, that the contrast is maximum when the tip is at the minima of the surface.

As the second instance, we perform computer simulations on the interaction of a tip in air with the evanescent field created under total internal reflection (TIR) of a focused light beam at an air-dielectric interface at subwavelength distance from the tip in the detection process of a Photon Scanning Tunneling Microscope (PSTM) [11]-[14]. This is of importance in Near Field Optics (NFO) [15] in order to assess the scope of measurements and to interpret the data. We predict the tunneling from the dielectric surface to the tip through the air. Also, the scanning tip is able to yield subwavelength information of details so small as the *Goos-Hänchen lateral shift* (GHS) of a narrow beam, namely, whose half width $W$ is a few wavelengths $\lambda$. Optical measurements of the GHS are scarce. Till now it has been detected by inference from multiple bouncing [16] or from laser cavity proving [17] for beams with $W >> \lambda$. The existing theories also apply to this regime, [18]-[20]. It has also been measured in microwave experiments [21], but the beams were of rather poor quality.

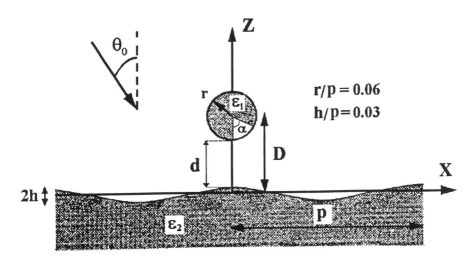

*Figure 1.* Scattering Geometry. The cylinder size and the surface parameters are not plotted at scale. The real surface is almost flat compared with the tip curvature. This configuration is chosen to excite a strong surface plasmon.

## 2. Subnanometric Variations of Field Enhancements between Tips and Surfaces

It has been suggested that the measured light emission is due to localized tip induced plasmons (TIP) modes from electromagnetic coupling [8]-[9]. Furthermore, it has been suggested by using a fitting model[5], that these TIP modes are sensitive to the tip-sample distance averaged over a surface area of the order of the TIP mode lateral extension. This implies that there is variation in the light intensity emitted at 600 nm when the tip surface distance changes by 0.1 nm [5]. Here we present the calculation for tips of Ag and W on an Ag or W surface, both being described by a frequency dependent complex dielectric constant. The calculation is done by a rigorous theory making use of the extinction theorem (ET) boundary condition[22]. It is shown that the plasmon field intensity, both on the tip and on the surface, varies by a 10% when the tip-surface distance changes by 0.1 nm, and also that this intensity distribution is concentrated in the tip-surface gap, a result that is in agreement with the experimental observation [5].

In order to establish the theory we proceed as follows: the physical system is shown in Fig. 1. It is two-dimensional, namely the geometry is in the $XZ$-plane, due to computer memory sake. This configuration contains, however, all the essential physics. The tip is simulated by a cylinder, whose axis is along $OY$, of dielectric permittivity $\epsilon_1$ and radius $r = 51.45$ nm, with its center at a distance $D$ from the plane $z = 0$, note that this model of tip is the 2-D version of assuming in 3-D the active part of the tip to be a sphere. A surface $z = S(x)$ of mean plane $z = 0$, separates vacuum, $z > S(x)$, from the sample, $z < S(x)$, with dielectric permittivity $\epsilon_2$. In order to excite a strong light-induced propagating surface plasmon on the grating[23], we choose a sample interface with a sinusoidal profile, $S(x) = h \cos(2\pi x/p)$, of period $p = 800$ nm and height $h = 24$ nm. A $p$-polarized (i.e. with the electric field $\mathbf{E}$ contained in the $XZ$-plane) plane electromagnetic wave of wavelenght $\lambda = 514.5$ nm is incident on the system from the vacuum side at an angle of incidence $\theta_0 = 24.5^0$. At this angle $\theta_0$, it is well known that there is absorption of the incident energy due to the excitation of the surface plasmon on the surface with the above parameters[24].

An exact numerical calculation of the electromagnetic interaction between the tip and the surface at subnanometer proximity is done next on using the ET for multiply connected scattering domains[25]. In this way, the equations governing the electric vector $\mathbf{E}$ for the system are:

$$\mathbf{E}(\mathbf{r}^<) = -\frac{1}{4\pi k_0^2 \epsilon} \nabla \times \nabla \times \int_{S_i} \left( \mathbf{E}_{in}(\mathbf{r}') \frac{\partial G^{(in)}(\mathbf{r}^<, \mathbf{r}')}{\partial \mathbf{n}} - G^{(in)}(\mathbf{r}^<, \mathbf{r}') \frac{\partial \mathbf{E}_{in}(\mathbf{r}')}{\partial \mathbf{n}} \right) ds,$$

$$(1)$$

$$0 = \mathbf{E}^{(i)}(\mathbf{r}^<) + \frac{1}{4\pi k_0^2} \nabla \times \nabla \times \sum_i \int_{S_i} \left( \mathbf{E}(\mathbf{r}') \frac{\partial G(\mathbf{r}^<, \mathbf{r}')}{\partial \mathbf{n}} - G(\mathbf{r}^<, \mathbf{r}') \frac{\partial \mathbf{E}(\mathbf{r}')}{\partial \mathbf{n}} \right) ds.$$

$$(2)$$

And:

$$0 = \frac{1}{4\pi k_0^2 \epsilon} \nabla \times \nabla \times \int_{S_i} \left( \mathbf{E}_{in}(\mathbf{r}') \frac{\partial G^{(in)}(\mathbf{r}^>, \mathbf{r}')}{\partial \mathbf{n}} - G^{(in)}(\mathbf{r}^>, \mathbf{r}') \frac{\partial \mathbf{E}_{in}(\mathbf{r}')}{\partial \mathbf{n}} \right) ds, \quad (3)$$

$$\mathbf{E}(\mathbf{r}^>) = \mathbf{E}^{(i)}(\mathbf{r}^>) + \frac{1}{4\pi k_0^2} \nabla \times \nabla \times \sum_i \int_{S_i} \left( \mathbf{E}(\mathbf{r}') \frac{\partial G(\mathbf{r}^>, \mathbf{r}')}{\partial \mathbf{n}} - G(\mathbf{r}^>, \mathbf{r}') \frac{\partial \mathbf{E}(\mathbf{r}')}{\partial \mathbf{n}} \right) ds.$$

$$(4)$$

In Eqs. (1)-(4) $S_i$ means the surface of either the cylinder ($i = 1$), or the grating ($i = 2$), $\mathbf{E}^{(i)}$ is the incident field and "$in$" denotes the limiting value

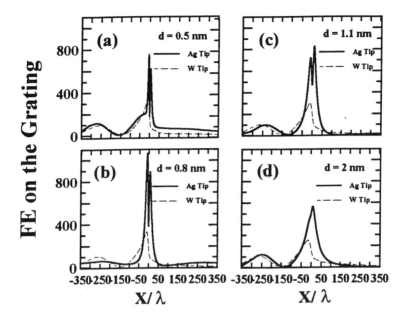

*Figure 2.* Normalized field enhancement (FE) on the grating surface. The sample is Ag with $\epsilon_2 = -11.04 + i0.33$. Grating parameters: $h = 24$ nm, $p = 800$ nm. The choice of these parameters are done in order to produce a very hight propagatig plasmon FE on the surface. The center of the cylinder is located at the abscisa $x = 0$. Solid line: Ag tip, $\epsilon_1 = -11.04 + i0.33$. Broken line: W tip, $\epsilon_1 = 4.5 + i18.77$. (a) $d = 0.5$ nm; (b) $d = 0.8$ nm; (c) $d = 1.1$ nm; (d) $d = 2$ nm

taken on the surface from inside the volume $V_i$. **n** denotes the unit outward normal; $G(\mathbf{r}, \mathbf{r}') = \pi i H_0^{(1)}(k_0 |\mathbf{r} - \mathbf{r}'|)$, $G^{(in)}(\mathbf{r}, \mathbf{r}') = \pi i H_0^{(1)}(\sqrt{\epsilon} k_0 |\mathbf{r} - \mathbf{r}'|)$. $H_0^{(1)}(.)$ being the zeroth-order Hankel function of the first kind, $k_0 = 2\pi/\lambda$. $\epsilon$ is $\epsilon_1$ on the cylinder or $\epsilon_2$ on the grating. $\mathbf{r}^<$ represents a point either inside the cylinder or in the sample, and $\mathbf{r}^>$ corresponds to a point in the air.

The radiation condition for the *scattered* fields, namely, the scattered fields angular spectrum representation contains *only* outgoing waves at infinity, plus the continuity conditions: $\mathbf{n} \times \left[ \mathbf{E}^{(in)}(\mathbf{r}^<) - \mathbf{E}(\mathbf{r}^>) \right] = 0$ and $\mathbf{n} \times \left[ \mathbf{H}^{(in)}(\mathbf{r}^<) - \mathbf{H}(\mathbf{r}^>) \right] = 0$ across the surfaces $S_i$ permit to find both $\mathbf{E}$ and $\partial \mathbf{E}/\partial \mathbf{n}$ or, equivalently, $\mathbf{H}$ and $\partial \mathbf{H}/\partial \mathbf{n}$, from either the pair of Eqs. (3) and (4) or, equivalently, from the pair of Eqs. (1) and (2), as both $\mathbf{r}^<$

and $\mathbf{r}^>$ tend to a point in $S_i$. Then, the total field in the air or in the tip is given by the second term of Eq. (4) or from Eq. (1) with $i = 2$, respectively.

Fig. 2 shows the normalized field enhancement (FE) defined as the ratio between the squared modulus of the total (i.e. scattered plus incident) field and the squared modulus of the incident field on the grating surface. The sample is Ag, $\epsilon_2 = -11.04 + i0.33$[26]. Two different instances of tips, namely a W tip with permittivity $\epsilon_1 = 4.5 + i18.77$[27] at $\lambda = 514.5$ nm and an Ag tip with $\epsilon_1 = -11.04 + i0.33$, have been addressed. The calculations have been done with the center of the cylinder located in front of a maximum of the surface profile[28] at different widths of the air gap between the tip and the sample. They are shown in Figs. 2(a-d), $d = 0.5, 0.8, 1.1$ and 2 nm, respectively. It is worth noting from these figures that there is a confinement together with an enhancement effect, induced by the presence of the tip, on the field intensity just inside the cavity formed by the tip and the sample. The magnitude of this FE is at least four times greater than the plasmon FE ($\simeq 100$), obtained for the surface profile without the existence of cylinder. Although the ratio between the FE in a system consisting of a cylinder in front of a plane interface and the plasmon FE in a plane surface without cylinder is the same as the ratio when the plane is substituted by a periodically modulated surface profile, the parameters of our surface have the important advantage of producing much larger background FE. At this point, it is important to remark an interesting feature in Fig. 2: the slightly decrease of the lateral width of the TIP mode as the distance tip-sample decreases. Furthemore, this width is in our calculation between 10 and 20 nanometers.

Also, the dependence of the FE on the tip dielectric permittivity is clear. The FE is always larger with the Ag tip than with the W tip. For the sake of clarity, in Fig. 3 we show the influence of the tip dielectric constant on the magnitude of the FE. In this case, the system consists of a W sample with the same profile as in the previous calculation. Although this surface does not support a propagating surface plasmon, a FE induced by the resonant system: tip-surface, still appears but being much smaller than for the Ag surface.

These results could be important for photoinduced tunneling current experiments[29]. It can be seen that while a tiny FE exists for a W sample with an Ag tip, there is a huge enhancement for an Ag surface with both an Ag or W tip (Fig. 2). Therefore, the photoinduced currents should be observable from Ag surfaces with a W or Ag tip because the difference in the FE between both cases is only a factor of 2. The FE is always defined by the large object, namely, the surface.

Fig. 4 shows the dependence of the FE peak versus the distance $d$ tip-sample for an Ag surface and the two tip materials considered above. Also,

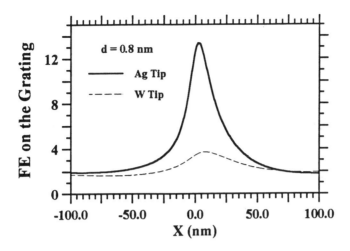

*Figure 3.* FE on the grating surface for a W sample, $\epsilon_2 = 4.5 + i18.77$ with a cylinder at $d = 0.8$ nm. Solid line: Ag tip, broken line: W tip.

the background plasmon FE that would exist without tip (solid line) and the one corresponding to a perfect conductor ($\epsilon = -\infty$) (broken line) are shown. There is a marked difference between our calculation and the dependence proportional to $1/d^2$ which appears in the perfect conductor model. It is worth noting that, while for the Ag tip there exits an optimal value of the tip-sample distance $d = 0.8$ nm, approximately, in which the FE has a maximum, for the W tip the FE has a monotonic increase behavior as the probe is approached to the interface. This fact can produce a contrast reversal in the photoemission images induced by a STM when the Ag tip (or another material with similar dielectric contants) is at a distance $d$ either less or greater than 0.8 nm. Nevertheless, at smaller distances caution should be taken because of changes in the Ag dielectric function due to the tip-sample proximity. However, it is clear for both tips that the FE located between tip and sample appreciably changes when the distance $d$ varies by less than 0.1 nm. This is in agreement with the experimental data reported in Ref. 5.

These effects can play an important role in other experiments [30] involving mixed interactions. The common feature in experiments of Refs. 5 and 30 is that they use different interactions for control and measure-

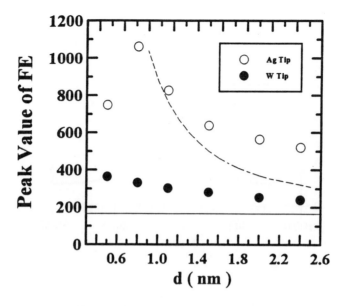

*Figure 4.* Peak value of the FE on the grating surface versus distance tip-sample $d$. Circles: Ag tip; Full circles: W tip. Solid line: background plasmon FE on the isolated surface (without tip). Broken line: FE value for the case of perfect metal approximation (see text for details).

ment, that is: either the tunneling current[5] or the interference between a reflected wave from a small tip and the incident light[30]. This has been shown to provide about 1 nm resolution[31]. Basically, the idea is the same in both experiments[5],[30], namely, a small tip modulates the background signal. This represents a breakthrough in near field optical microscopy.

## 3. Detection of Goos-Hänchen shifts

The tip is then modelled by a cylinder, whose axis is along $OY$, of permittivity $\epsilon_2$ and radius $a$, with its center at distance $d + a$ from the interface. An $s$-polarized focused Gaussian beam of half width $W$ is incident on the surface from the dielectric side. (For brevity, only $s$-waves are addressed). We have used the method shown in Ref. 1 and 25. The calculations exhibit unitarity and reciprocity within 99%. We propose here a detection of sub-wavelength Goos-Hänchen shifts for highly focused beams from NFO measurements. In our numerical experiments, and for computer memory sake, the system is two-dimensional, namely, the geometry is in the $XZ$-plane,

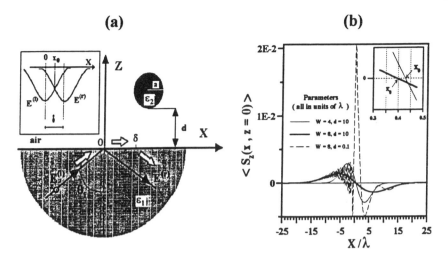

*Figure 5.* (a): Scattering geometry. The inset shows the field distribution of the incident beam ($\mathbf{E}^{(i)}$) and of the reflected shifted beam ($\mathbf{E}^{(r)}$), respectively. The white arrows in the main figure show the 'bending' effect of the Poynting vector. (The two plots are not shown in the same scale). (b): Density of energy flux across the dielectric interface $z = 0$ ($z$-component of the time average Poynting vector, $< S_z(x, z = 0) >$) in arbitrary units for cases in which $d \gg \lambda$. The inset shows a magnification over the interval $x/\lambda = [0.3, 0.5]$. $a = 0.5\lambda$, $\epsilon_1 = \epsilon_2 = 2.12$, $\theta_i = 50°$. Thick solid line: $W = 4\lambda$, $d = 10\lambda$. Thin solid line: $W = 8\lambda$, $d = 10\lambda$. Broken line: $W = 8\lambda$, $d = 0.1\lambda$

(Fig. 5(a)). The dielectric below the plane interface $z = 0$ has permittivity $\epsilon_1$.

Fig. 5(b) shows the time average Poynting vector across the plane $z = 0$: $< S_z(x, z = 0) >$ for beams of half width $W = 8\lambda$ and $4\lambda$, incident at an angle $\theta_i = 50°$, from the dielectric side. $\epsilon_1 = \epsilon_2 = 2.12$. The tip, centered at $x = 0$, is considered with $a = 0.5\lambda$, and $d$ either $d = 0.1\lambda$ or $d = 10\lambda$. It is observed that the distance $d = 10\lambda$ is large enough so that there is no perturbation of the cylinder on the energy flow that would exist in the absence of tip. *The energy flows to and fro. An amount enters into the air (namely, it is positive for $x/\lambda < x_0$) giving rise to an evanescent field there, and then it goes back into the dielectric (i.e., it is negative for $x/\lambda > x_0$). The net flow $\Phi = \int_{-L/2}^{L/2} < S_z(x', z = 0) > dx'$ is zero* [32]. The sign of the local value of $< S_z >$ changes in $OX$ about the point $x = x_0$ due to the *lateral shift* along this direction between the incident and the

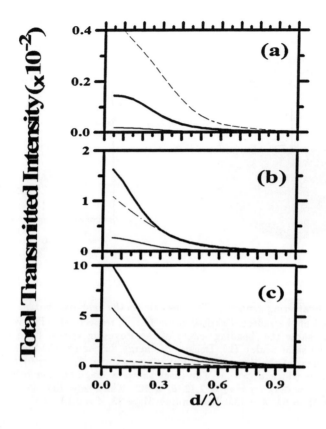

*Figure 6.* Normalized total transmitted intensity at infinity versus distance $d$ cylinder-plane. $\theta_i = 50°$, $\epsilon_1 = 2.12$, $W = 8\lambda$. Thick solid line: $\epsilon_2 = 3.8$; thin solid line: $\epsilon_2 = 2.12$; broken line: $\epsilon_2 = -9.89 + i1.05$.

reflected beam. Hence, *the value $x_0$ of the abscisa $x/\lambda$ is precisely half the Goos-Hänchen shift*: $\delta = 0.80\lambda$ and $\delta = 0.86\lambda$ for $W = 8\lambda$ and $W = 4\lambda$, respectively, as it is seen from the inset of Fig.2, (see also Fig. 5(a)). As $W$ decreases in this range, this shift increases. (This variation of $< S_z >$ illustrates how the direction of the Poynting vector $< \mathbf{S}(x, z) >$ changes locally: initially, at $x = 0$, pointing towards $z > 0$, then progressively "bending" thus becoming parallel to the interface at the intermediate point $x = x_0$, until it reemerges into $z < 0$ at $x = \delta$, [33], [34]).

When this rather large tip is approached to the interface, (i.e. $d = 0.1\lambda$), this flow distribution is perturbed by the coupling of the evanescent field with propagating waves through the cylinder surface and there is a net

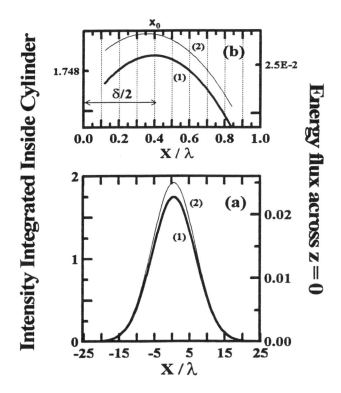

*Figure 7.* (a): Integrated intensity inside the cylinder at constant height $z = d = 0.15\lambda$ (curve 1), and total energy flux $\Phi$ entering $z > 0$ (curve 2), versus lateral position of the cylinder. $W = 8\lambda$, $\theta_i = 50°$, $\epsilon_1 = \epsilon_2 = 2.12$, $a = 0.1\lambda$. (b): Same as (a) magnified in the interval $x/\lambda = [0, 1]$

flow of energy into the air, ($\Phi = 0.35$ for $W = 8\lambda$). Fig. 6 shows the total intensity transmitted to infinity through this coupling for nine different cylinders of radii: $a = 0.5\lambda$, $0.1\lambda$, and $0.05\lambda$, and permittivities: $\epsilon_2 = 3.8$, 2.12 and $-9.89 + i1.05$ (gold at $\lambda = 652.6$ $nm$), showing the incresing conversion of the evanescent waves into waves propagating to the far zone as the distance $d$ decreases. (Due to the scattering process in the presence of the cylinder and the propagation to the far zone, these curves do not fit a negative exponential). Also, this coupling increases as the radius $a$ of the tip becomes larger as then it has a wider surface to produce this matching. Nevertheless, it is worth noting that for very tiny tips, a metallic material yields a larger signal than a dielectric one (compare Fig 6(a) v.s. 6(b) and 6(c)): whereas for larger tips the absorption effects produced by the metal obliterate the coupling, for smaller $a$ the tinier volume of the tip does not produce so much absorption and the surface effect predominates. We observe that tips with radii smaller than $0.15\lambda$ and $\epsilon \leq 2.12$ do not appreciably produce perturbation of the structure of the near field while they produce the coupling, thus they are little invasive. We have simulated the field collected in one of such tips .

On the other hand, Figs. 7(a) and (b) show: (1) The intensity inside the same cylinder, integrated along the tip diameter, and (2) The total time average energy flow $\Phi$ entering into $z > 0$ across the plane $z = 0$, both quantities now being calculated as this cylinder scans with its center moving parallel to the plane along the line $z = d = 0.15\lambda$. The curve (1) shows the detection by the tip of the evanescent Gaussian beam in the region $z > 0$. Having the incident beam a Gaussian section on the plane $z = 0$ of width $W = 8\lambda$, centered at $x = 0$ , the Gaussian envelope detected by the tip corresponds, by continuity of the fields through the interface, to the sum of this incident field plus the reflected one shifted by the GHS $\delta$. Since $\delta$ is small, this sum is very approximately another Gaussian profile centered at $x_0$, namely, at half distance between the peaks of the Gaussian envelopes of both fields. As a consequence, $x_0$ represents half the Goos-Hänchen lateral shift $\delta = 0.80\lambda$ between the incident and the reflected beam. (See Fig. 5(b)).

Therefore, and in conclusion, a scanning with a dielectric tip of radius not larger than $0.15\lambda$ at subwavelength distance from the interface can exhibit shifts smaller than the wavelength, suffered by very narrow focused beams under total internal reflection, once the position of the incident beam is fixed. Other surface wave features of subwavelength size could equally be envisaged. Experiments can be done on the basis of this analysis.

This research has been supported by the Comision Interministerial de Ciencia y Tecnologia under grant PB 92-0081. A.M. acknowledges a scholarship from Comunidad Autonoma de Madrid.

# References

1. A. Madrazo and M. Nieto-Vesperinas, J. Opt. Soc. Am. (in press)
2. R. Carminati and J.J. Greffet, Opt. Comm. **116**, 316 (1995)
3. L. Novotny, D.W. Pohl and P. Regli, J. Opt. Soc. Am. A **11**, 1768 (1994)
4. C. Girard and A. Dereaux, Phys. Rev. B **49**, 11344 (1994).
5. R. Berndt, R. Gaisch, W. D. Schneider, J. K. Gimzewski, B. Reihl , R. R. Schlittler and M. Tschudy, Phys. Rev. Lett. **74**, 102 (1995).
6. J. Lambe and S.L. McCarthy, Phys. Rev. Lett. **37**, 923 (1976).
7. J.H. Coombs, J.K. Gimzewski, B. Reihl, J.K. Sass and R.R. Schlittler, J. Microscopy **152**, 325 (1988).
8. B.N.J. Persson and A. Baratoff, Phys. Rev. Lett. **68**, 3224 (1992).
9. R. Berndt, J.K. Gimzewski and P. Johansson, Phys. Rev. Lett. **71**, 3493 (1993).
10. M. Tsukuda, T. Shimizu and K. Kobayashi, Ultramicroscopy **42-44**, 360 (1992).
11. D. Courjon, K. Serayeddine and M. Spajer, Optics Comm. **71**, 23 (1989); D. Courjon, C. Bainier and M. Spajer, J. Vac. Sci. Technol. B **10**, 2436 (1989).
12. P. Dawson, F. de Fornel and J.P. Goudonnet, Phys. Rev. Lett. **72**, 2979 (1994); F. de Fornel, P.M. Adam, L. Salomon, J.P. Goudonnet and P. Guerin, Opt. Lett. **14**, 1082 (1994).
13. J. J. Greffet, A. Sentenac and R. Carminati, Opt. Comm. **116**, 20 (1995); R. Carminati and J. J. Greffet, Opt. Comm. **116**, 316 (1995).
14. N. Garcia and M. Nieto-Vesperinas, Opt. Lett. **24**, 2090-2092, (1993); R. Carminati, A. Madrazo and M. Nieto-Vesperinas, Opt. Comm. **111**, 26-33 (1994); N. Garcia and M. Nieto-Vesperinas, Opt. Lett. **20**, 949 (1994).
15. D. W. Pohl, "Scanning Near Field Optical Microscopy (SNOM)" in *Advances in Optical and Electron Microscopy*, J. R. Sheppard and T. Mulvey, eds., (Academic Press, New York, 1990), p. 243; D. W. Pohl and D. Courjon, eds, *Near Field Optics*, NATO ASI Series, Vol. 242, Kluwer, Dordrecht, 1993.
16. F. Goos and H. Hänchen, Ann. Physik **1**, 333 (1947); F. Goos and H. Hänchen, Ann. Physik **5**, 251 (1951); C. Imbert and Y. Levy, Nouv. Rev. Optique **6**, 285 (1975).
17. F. Bretenaker, A.L Floch and L. Dutriaux, Phys. Rev. Lett. **68**, 931 (1992).
18. K. Artmann, Ann. Physik **2**, 87 (1948).
19. H. K. V. Lotsch, J. Opt. Soc. Am. **58**, 551 (1968).
20. H. R. Horowitz and T. Tamir, J. Opt. Soc. Am. **61**, 586 (1971); F. Falco and T. Tamir, J. Opt. Soc. Am. A **7**, 185 (1990); J. J. Greffet and C. Baylard, Opt. Comm. **93**, 271 (1992).
21. J. J. Crowan and B. Anicin, J. Opt. Soc. Am. **67**, 1307 (1977); M. Wong, G. E. Reesor and L. A. A. Read, Can. J. Phys. **55**, 1061 (1977).
22. M. Nieto-Vesperinas, *Scattering and Diffraction in Physical Optics* (Wiley, New York, 1991). Chapter 1.
23. N. Garcia, G. Diaz, J.J. Saenz and C. Ocal, Surf. Sci. **143**, 342 (1984).
24. N. Garcia, Opt. Comm. **45**, 307 (1983).
25. A. Madrazo and M. Nieto-Vesperinas, J. Opt. Soc. Am. A **12**, 1298 (1995).
26. P.B. Johnson and R.W. Christy, Phys. Rev. B **6**, 4370 (1972)
27. David W. Lynch and W.R. Hunter, in *Handbook of Optical Constants of Solids*, edited by E.D. Palik (Academic Press, New York, 1985), p. 366.
28. It is important to remark the fact that the results reported here do not dependent on the tip lateral position above the corrugated surface. The large range 800 nm surface periodicity should not be mixed with the sort 0.8 nm atomic periodicity of the experiments in Ref. 1. The nanometer scale corrugation addressed in our calculation is only used in order to produce a very hight efficient coupling between the incident electromagnetic energy and the propagating surface plasmon. The atomic resolution comes out as a consequence of the variations of the FE when the STM is operating at constant current mode.

40

29. C. Baur, B. Koslowski, R. Möller and K. Dransfeld, in *Near Field Optics*, edited by D.W. Pohl and D. Courjon, NATO Advanced Studies Institutes Series E (Kluwer, Dordrecht, The Netherlands , 1993), Vol. 241, pp 325-331.
30. F. Zenhausern, Y. Martin and Wickramasinghe, Science **269**, 1083 (1995).
31. N. Garcia and M. Nieto-Vesperinas, Appl. Phys. Lett. **66**, 3399 (1995).
32. M. Born and E. Wolf, *Principles of Optics*, Pergamon Press, Oxford, 1975, Section 1.5.4.
33. J. Pitch, Ann. Physik **3**, 433 (1929); J. Pitch, Optik **12**, 41 (1955).
34. R. H. Renard, J. Opt. Soc. Am. **54**, 1190 (1964).

# A NUMERICAL STUDY OF A MODEL NEAR-FIELD OPTICAL MICROSCOPE

A. A. MARADUDIN

*Department of Physics and Astronomy*
*and Institute for Surface and Interface Science*
*University of California*
*Irvine, CA 92717, U.S.A.*

A. MENDOZA-SUÁREZ AND E. R. MÉNDEZ

*División de Física Aplicada*
*Centro de Investigación Científica*
*y de Educación Superior de Ensenada*
*Apdo Postal 2732, Ensenada B.C. 22800, México*

AND

M. NIETO-VESPERINAS

*Instituto de Ciencia de Materiales, CSIC*
*Facultad de Ciencias C-III, Universidad Autónoma*
*Campus de Cantoblanco*
*E-28049 Madrid, Spain*

**Abstract.** We study a two-dimensional model of the scattering of p- and s-polarized light, emitted by a coated tapered glass fiber, from a metal surface with a topographic or an optical defect. We calculate the intensity of the scattered electromagnetic field at the center of the bottom of the fiber, and the integrated intensity of the field scattered back through the fiber, as it is moved at constant height above the perturbed metal surface. The position and width of surface defects can be determined from these calculations with subwavelength resolution, although the intensities bear no simple relation to the surface scanned.

41

*M. Nieto-Vesperinas and N. García (eds.), Optics at the Nanometer Scale* 41–61.
© *1996 Kluwer Academic Publishers. Printed in the Netherlands.*

# 1. Introduction

In near-field optical microscopy an electromagnetic field scattered from a surface, or transmitted through it, is detected by a sharpened, often coated, optical fiber, called a tip, that is brought to within a small fraction of the wavelength of the incident electromagnetic field from the surface, and is scanned along it. The source of the electromagnetic field can be the optical fiber itself, in the former case, or a source external to it, in both cases. In a theoretical investigation of the characterization of the topographical and/or optical properties of surfaces by this technique, two related problems arise. The first is the formation of an image, i.e. the determination of the field measured by the tip, while the second is the relation of the image to the topography of the surface and to its optical properties. In this paper we describe a computational approach to the solution of the former problem–to the formation of an image–which is based on the Green's function surface integral equation approach that has proved to be so successful in theoretical studies of the scattering of electromagnetic waves from randomly rough surfaces[1, 2]. Our aim is to explore the extent to which, and the manner in which, topographic features on the surface, with dimensions much smaller than the wavelength of the incident electromagnetic field, show up in the calculated images. We emphasize, however, that the images calculated in this paper bear no simple relation to the topography of the surface nor to its optical properties. Approaches to obtaining topographical information about the surface scanned by the optical fiber from the images thus obtained have been presented by several authors[3, 4, 5]. Some of this work[3, 4] has been based on the assumption that the optical fiber is a passive probe, i.e. that its presence does not perturb the electromagnetic field being measured by it. In other work[5] it has been shown that, at least for dielectric surfaces, the effect of the tip on the field being measured is fairly independent of the surface under examination, so that the topography of that surface can be obtained from the image through the use of a transfer function. Our own work on this problem will be published elsewhere[6].

Underlying the present calculations are the beliefs that in any theory of near-field optical microscopy the presence of the optical fiber must be taken into account in the calculation of the electromagnetic field scattered from the surface being imaged, regardless of the source of the incident field, until the conditions under which the presence of the fiber does not perturb the field being measured by it are understood, and that the ability to calculate the scattered field accurately in the presence of the fiber is crucial to achieving that understanding.

The outline of this paper is the following. In Section 2 we describe the model of a near-field optical microscope on which the calculations described in this paper are based, and define the two types of images that are formed by it. The surface structures imaged by this microscope are presented in Section 3. The Green's function surface integral equation approach to the calculation of the images of these surface structures that are recorded as the microscope is translated parallel to the surface on which they are situated, is outlined in Section 4. The results obtained by this approach are presented in Section 5. A discussion of these results, and the conclusions drawn from them in Section 6, end this paper.

## 2. The Model Microscope

The model near-field microscope to be studied in this work is depicted schematically in Fig. 1. It consists of a tapered glass fiber that projects downward from a semi-infinite cladding of the same glass toward the surface of a semi-infinite metal substrate that is to be imaged. The metal surface is assumed to be planar except for topographical or optical perturbations which change its height or its dielectric properties, respectively, over a limited region of the surface. An example of a topographical perturbation is depicted in Fig. 1. The system is translationally invariant along the $x_2$-axis, i.e. in the direction normal to the plane of the figure. The tapered walls of the glass fiber, and the remainder of the glass-vacuum interface, except for the horizontal portion of the bottom of the fiber, are assumed to be coated with an infinitesimally thin perfectly conducting film. The presence of this film forces the tangential component of the electric field and the normal component of the magnetic field in the system to vanish along the coated glass-vacuum interface. The dielectric constant of the glass fiber and the glass cladding in the present work is chosen to be $\epsilon = 2.25$. The region between the glass fiber and the metal surface is vacuum.

An electromagnetic plane wave of p- or s-polarization, with the $x_1x_3$-plane the plane of incidence, is incident normally on this structure from the glass side. The wavelength of the incident electromagnetic field will be taken to be $\lambda = 0.65\mu m$ in all the calculations whose results are described in this paper. A certain portion of this incident field passes through the glass fiber and impinges on the metal surface, from which it is scattered. A part of the scattered field is transmitted through the fiber back into the glass region above it.

When the microscope is in operation, the glass fiber is translated parallel to the metal surface that is being imaged. The horizontal portion of the bottom of the fiber is kept at a constant height above the planar metal surface that is perturbed by the localized topographical or optical discontinuities. At each point of this horizontal scan of the surface an image is formed by the glass fiber. This image is defined in two different ways. The

44

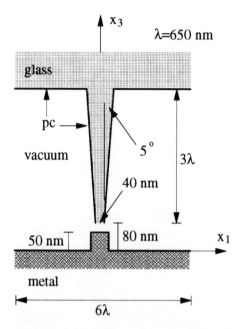

*Figure 1.* A schematic depiction of the model near-field microscope studied in the present work. It is shown imaging a rectangular ridge on a metal surface. The heavy lines denote a perfectly conducting film separating the glass region from the vacuum region. A plane electromagnetic wave is incident normally on this structure from the vacuum side. Part of the electromagnetic field scattered by the metal surface is transmitted through the fiber back into the glass region above it.

first definition of an image is that it is the squared modulus of the single nonzero component of the scattered magnetic (electric) field at the center of the uncoated, horizontal bottom of the fiber, in the case of illumination by a p- (s-)polarized electromagnetic field. The second definition of image is that it is the squared modulus of the amplitude of the field scattered back into the glass cladding through the fiber, integrated over all scattering angles, viz. it is given by

$$\int_{-\frac{\pi}{2}}^{\frac{\pi}{2}} |r_{p,s}(\theta_s)|^2 d\theta_s,$$

where explicit expressions for the scattering amplitudes $r_{p,s}(\theta_s)$ for p- and s-polarized light, respectively, are given by Eqs. (3.30) and (4.16) of Ref. 7. In what follows these two types of image will be referred to as the near-field and far-field images, respectively. We note, however, that while the far-field images represent measurable quantities, the near-field images are somewhat artificial, given the experimental difficulty of separating the scattered field from the total field at the center of the uncoated bottom of the fiber. We have considered this definition of the image because in the future its use could lead to a simplification of the inversion problem, and also because it is analogous to the definition of image used by other authors[3, 4, 5].

## 3. The Surface Structures Imaged

In this work the near-field optical microscope described in the preceding section will be used to image the six different surface structures that are depicted in Fig. 2. Each of these surface structures is a topographical or optical perturbation of an otherwise planar, homogeneous metal surface. They include (a) a metal ridge of square cross section formed from the same metal as the substrate on which it rests; the ridge is $0.05\mu m$ wide and $0.05\mu m$ high; (b) a metal ridge of rectangular cross section formed from the same metal as the substrate; the width of the ridge is $0.025\mu m$, while its height is again $0.05\mu m$; (c) a pair of rectangular ridges, each with the dimensions given in (b) above, formed from the same metal as the substrate, whose inner walls are separated by $0.05\mu m$; (d) a pair of rectangular ridges, each with the dimensions given in (b) above, formed from the same metal as the substrate, whose inner walls are now separated by $\lambda/2 = 0.325\mu m$; (e) a step of height $0.05\mu m$ formed from the same metal as the substrate. The linear dimensions of these five surface structures are thus significantly smaller than the wavelength of the incident electromagnetic wave, and are comparable with the width of the horizontal bottom of the fiber ($0.04\mu m$). The dielectric function of the metal from which these five surface structures are formed is $\epsilon(\omega) = -11.36 + i0.96$, which is the dielectric function of gold at a wavelength $\lambda = 0.65\mu m$. The sixth surface to be imaged is (f) a planar surface that bounds an optical discontinuity: the quadrant $x_1 < 0, x_3 < 0$ is a metal that is characterized by a dielectric function $\epsilon_1(\omega) = -11.36 + i0.96$, while the quadrant $x_1 > 0, x_3 < 0$ is a different metal that is characterized by a dielectric function $\epsilon_2(\omega) = -17.15 + i5.88$. These are the dielectric functions of gold and silver, respectively, at a wavelength $\lambda = 0.65\mu m$.

In all determinations of the images of the six surface structures just described, the horizontal bottom of the glass fiber is maintained at a height of $0.08 \mu m$ above the planar metal surface perturbed by these structures.

## 4. The Method of Calculation

Because the system depicted in Fig. 1 is invariant in the $x_2$-direction, and the plane of incidence of the electromagnetic field is the $x_1 x_3$-plane, there is no cross-polarized scattering in it. It is therefore convenient to work with the single nonzero component of the magnetic field in the system, $H_2(x_1, x_3|\omega)$, when the incident electromagnetic field is p-polarized, and with the single

46

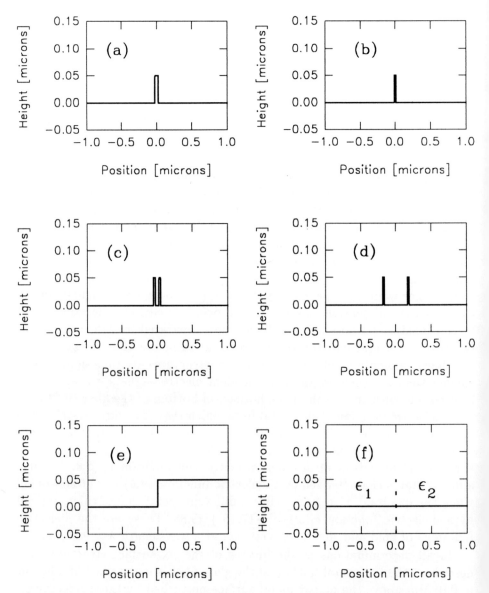

*Figure 2.* The six surface structures whose images are calculated in this paper. (a) A ridge of square cross section $0.05\mu m$ wide and $0.05\mu m$ high, formed from the same metal as the substrate on which it rests; (b) a ridge of rectangular cross section $0.025\mu m$ wide and $0.05\mu m$ high formed from the same metal as the substrate; (c) two rectangular ridges, each with the dimensions given in (b) above, formed from the same metal as the substrate, whose inner walls are separated by $0.05\mu m$; (d) two rectangular ridges, each with the dimensions given in (b) above, formed from the same metal as the substrate, whose inner walls are now separated by $\lambda/2 = 0.325\mu m$; (e) a step of height $0.05\mu m$ formed from the same metal as the substrate; (f) an optical discontinuity in which the quadrant $x_1 < 0, x_3 < 0$, is a metal whose dielectric function is $\epsilon_1(\omega)$, while the quadrant $x_1 > 0, x_3 < 0$ is a metal whose dielectric function is $\epsilon_2(\omega)$.

nonzero component of the electric field in the system, $E_2(x_1, x_3|\omega)$, when it is s-polarized. Each of these field components satisfies a Helmholtz equation in each of the regions, glass, vacuum, and metal, that comprise the system depicted in Fig. 1,

$$\left(\frac{\partial^2}{\partial x_1^2} + \frac{\partial^2}{\partial x_3^2} + \epsilon\frac{\omega^2}{c^2}\right) F(x_1, x_3|\omega) = 0. \tag{4.1}$$

In Eq. (4.1) $F(x_1, x_3)$ is either $H_2(x_1, x_3|\omega)$ or $E_2(x_1, x_3|\omega)$, and $\epsilon$ denotes the dielectric constant of the medium in which this equation is being solved. We have assumed a time dependence of the form $\exp(-i\omega t)$ for each of these field components, and have suppressed explicit reference to this factor.

Also needed in the discussion that follows is the two-dimensional, scalar Green's function $G_\epsilon(x_1, x_3|x_1', x_3')$ for the Helmholtz operator in Eq. (4.1), which satisfies the equation

$$\left(\frac{\partial^2}{\partial x_1^2} + \frac{\partial^2}{\partial x_3^2} + \epsilon\frac{\omega^2}{c^2}\right) G_\epsilon(x_1, x_3|x_1', x_3') = -4\pi\delta(x_1 - x_1')\delta(x_3 - x_3'), \tag{4.2}$$

subject to outgoing boundary conditions at infinity. It has the representation

$$G_\epsilon(x_1, x_3|x_1', x_3') = i\pi H_0^{(1)}\left(\sqrt{\epsilon}\frac{\omega}{c}[(x_1 - x_1')^2 + (x_3 - x_3')^2]^{\frac{1}{2}}\right), \tag{4.3}$$

where $Re\sqrt{\epsilon} > 0$, $Im\sqrt{\epsilon} > 0$, and where $H_0^{(1)}(z)$ is a zero-order Hankel function of the first kind.

To calculate the images of the objects depicted in Fig. 2 we apply Green's second integral identity in the plane[8], which in the present case can be stated as

$$\int_A \left\{ u(x_1, x_3)\left(\frac{\partial^2}{\partial x_1^2} + \frac{\partial^2}{\partial x_3^2} + \epsilon\frac{\omega^2}{c^2}\right) v(x_1, x_3) \right.$$

$$\left. -v(x_1, x_3)\left(\frac{\partial^2}{\partial x_1^2} + \frac{\partial^2}{\partial x_3^2} + \epsilon\frac{\omega^2}{c^2}\right) u(x_1, x_3) \right\} dx_1 dx_3$$

$$= \int_C \left\{ u(x_1, x_3)\frac{\partial}{\partial\nu}v(x_1, x_3) - v(x_1, x_3)\frac{\partial}{\partial\nu}u(x_1, x_3) \right\} ds. \tag{4.4}$$

In Eq. (4.4) $u(x_1, x_3)$ and $v(x_1, x_3)$ are any two functions of $x_1$ and $x_3$ which together with their partial derivatives are continuous inside and on the boundary of a region A of the $x_1 x_3$-plane bounded by a closed curve C; $\partial/\partial\nu$

denotes a derivative along the normal to the curve C at each point directed away from the region A; and $ds$ is the element of arc length along the curve C. If we set $u(x_1, x_3) = F(x_1, x_3\omega)$, and $v(x_1, x_3) = G_\epsilon(x_1, x_3|x_1', x_3')$, in Eq. (4.4), we can express the field component $F(x_1, x_3|\omega)$ at any point inside the region A in the form

$$F(x_1, x_3|\omega) = F^{(i)}(x_1, x_3|\omega)\theta_A(x_1, x_3)$$

$$-\frac{1}{4\pi}\int_C \left\{\left[\frac{\partial}{\partial\nu'}G_\epsilon(x_1, x_3|x_1', x_3')\right]F(x_1', x_3'|\omega)\right.$$

$$\left. - [G_\epsilon(x_1, x_3|x_1', x_3')]\frac{\partial}{\partial\nu'}F(x_1', x_3'|\omega)\right\}ds', \qquad (4.5)$$

where $F^{(i)}(x_1, x_3|\omega)$ is the incident field , a plane wave in our calculations, and $\theta_A(x_1, x_3)$ is the characteristic function of the region A, i.e. it equals unity if the point $(x_1, x_3)$ is inside A and is zero otherwise. If we apply Eq. (4.5) in turn to the glass region ($F^{(i)}(x_1, x_3|\omega)$ is nonzero only inside this region), to the vacuum region, and metal region, closing each at infinity as needed, we obtain the electromagnetic field $F(x_1, x_3|\omega)$ at any point inside each of these regions in terms of integrals along the surfaces (interfaces) bounding them, in which the values of $F(x_1, x_3|\omega)$ and its normal derivative evaluated on each surface appear. By letting the point of observation approach a point on each bounding surface from both sides, and using the boundary conditions satisfied on each surface by $F(x_1, x_3|\omega)$ and its normal derivative, one obtains a set of coupled, inhomogeneous integral equations for the independent boundary values of $F(x_1, x_3|\omega)$ and $\partial F(x_1, x_3|\omega)/\partial\nu$. This system of coupled integral equations can be converted into a set of inhomogeneous matrix equations by replacing the integrations in them by summations through the use of a numerical integration scheme. This system of matrix equations can be solved by a linear equation solver algorithm if it is not too large, or by an iterative approach, such as the conjugate-gradient method, if it is large. Once the boundary values of the field and its normal derivative have been calculated, the magnetic or electric field at any point of the system can be calculated from Eq. (4.5), and hence the images defined above.

In fact, in the present work the computational program just outlined was carried out with some simplifying modifications. In most calculations of the scattering of p- or s-polarized light from (say) a one-dimensional rough metal surface defined by the equation $x_3 = \zeta(x_1)$, the surface profile function $\zeta(x_1)$ is assumed to be a single-valued function of $x_1$. In this case, one can transform the integrals along the surface $x_3 = \zeta(x_1)$ into integrals along the $x_1$-axis[7]. However, in recent work by Mendoza-Suárez and Méndez[9] a parametrization of the equation defining the surface was devised that makes integration along the profile, even when it is re-entrant, i.e. multi-valued, straightforward. This approach was used in the present calculations, and increased the accuracy of the results, because a large number of equally spaced discretization points along the tapered sides of the glass fiber could be used. In addition, the use of this approach made it possible to image objects with vertical sides, which is not possible if integration along the surface $x_3 = \zeta(x_1)$ is replaced by integration along the $x_1$-axis.

Moreover, the application of Green's second integral identity to the region occupied by the metal was avoided by the use of an impedance boundary condition at the metal-vacuum interface. The impedance boundary condition used was the one for a single-valued curved metal-vacuum interface derived by García-Molina *et al.*[10], modified recently by Mendoza-Suárez and Méndez[11] through the use of their surface parametrization approach[9] to be applicable to multi-valued surfaces, including surfaces with vertical segments. This step significantly reduced the size of the system of matrix equations for the boundary values of $F(x_1, x_3|\omega)$ and its normal derivative that had to be solved.

The purely formal derivation of the integral equations for the boundary values of $F(x_1, x_3|\omega)$ and its normal derivative assumes that the system being studied is of infinite extent in the $x_1$-direction. In our numerical calculations we could work only with systems of finite extent along the $x_1$-axis. As we have indicated in Fig. 1, the system we used covered a length of only $6\lambda$ along the $x_1$-axis. In converting integrals along the surfaces in it into sums, the glass-vacuum interface was subdivided into 370 equally spaced points along it, separated by a constant distance $\Delta s = \lambda/32.5 = 0.02\mu m$. The vacuum-metal interface was subdivided into approximately 198 equally spaced points along it, with the precise number depending on the particular object being imaged. The separation between consecutive points on this interface was also $\Delta s = 0.02\mu m$. Because of the boundary conditions on $F(x_1, x_3|\omega)$ and $\partial F(x_1, x_3|\omega)/\partial \nu$ along the coated glass-vacuum interface, only two source functions are associated with this interface, one associated with its upper side and the other with its lower side. Because of the impedance boundary condition assumed on the vacuum-metal interface, only one source function is associated with this interface. The size of the matrix equation that had to be solved for the values of these source functions was therefore $2(370) + 198 = 938$. In the present work this equation was solved by a linear equation solver algorithm for each position of the glass fiber relative to the object being imaged.

Numerical experiments showed that in simulating the scanning of the metal surface by the glass fiber, while retaining a constant length of both the upper and lower surfaces, it was preferable to displace the upper surface

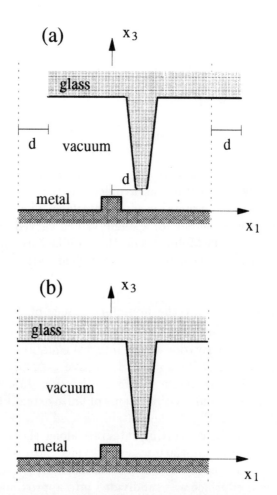

*Figure 3.* Two ways of modeling the displacement of the glass fiber over a rough metal surface while keeping the lengths of the upper and lower surfaces constant: (a) the upper surface is displaced rigidly to the right; (b) the fiber alone is displaced to the right.

rigidly, as in Fig. 3(a), rather than to displace only the fiber, as in Fig. 3(b). In the latter case the calculated images in p-polarization displayed spurious fine oscillations as functions of the position of the center of the fiber, which were attributed to numerical problems that arose in the use of that approach. Therefore, the approach to scanning depicted in Fig. 3(a) was the one used in the present calculations.

## 5. Results

To help interpret the images we have obtained we present in Figs. 4 and 5 gray level maps of the logarithm of the modulus of the scattered field

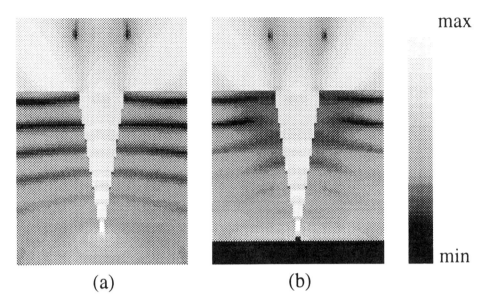

max

min

(a)                              (b)

*Figure 4.* Gray level maps of the logarithm of the modulus of the magnetic field amplitude in the vacuum region around the glass fiber when the system depicted in Fig. 1 is illuminated by a p-polarized electromagnetic field (a) in the absence of the metal substrate below it, and (b) in the presence of the metal structure depicted in Fig. 2(a) directly below the glass fiber.

amplitude around the glass fiber, both in the absence of any object to be imaged and in the presence of an object. In the latter case, the object used is the metal ridge of square cross section on an otherwise planar surface depicted in Fig. 2(a). Each figure represents an area that is approximately $2\lambda$ long in the $x_1$-direction, and $5\lambda$ long in the $x_3$-direction.

In Fig. 4(a) we present a gray level map of the logarithm of the modulus of the amplitude of the single, nonzero component of the magnetic field around the glass fiber when the system depicted in Fig. 1 is illuminated by a p-polarized electromagnetic field, in the absence of the metal substrate below the bottom of the fiber. It is seen that despite the narrow width of the fiber at its lower end, there is a fairly strong magnetic field at and below the bottom of the fiber. This is due to the fact that in a planar waveguide consisting of a dielectric layer sandwiched between perfectly conducting media, there is no cutoff for the propagation of p-polarized electromagnetic waves through it. The slight tapering of the walls of the coated glass fiber in the present case does not invalidate this result. Consequently some portion of the incident electromagnetic field is able to propagate through the fiber and emerge from the lower end. This field extends into the vacuum region above the bottom of the fiber on either side of it, and leads to a striated intensity pattern when it interferes with the field reflected from the coated, perfectly conducting, horizontal portion of the glass-vacuum interface. There is also a strong magnetic field at the center of the upper end of the glass fiber, where it merges with the glass cladding, and in scattering directions corresponding to scattering angles of $\theta_s \cong \pm 68°$. When the incident light impinges on an object rather than on empty space directly below it (Fig. 4(b)), there is a fairly strong magnetic field in the region between the bottom of the tip and the object, which spreads out along the metal surface on both sides of the object, and is presumably associated with the surface plasmon polariton excited through the interaction of the incident electromagnetic field with the object. The scattering of the field from the object decreases its value in the vicinity of the coated, perfectly conducting, horizontal portion of the glass-vacuum interface, but does not make a significant quantitative change in the field distribution in the glass cladding above the fiber. In the results presented in both Figs. 4(a) and 4(b) the field strength is seen to have an oscillatory variation along the axis of the glass fiber. This is presumably caused by the interference between the scattered light traveling down the fiber and the light reflected up the fiber from the uncoated glass-vacuum interface at its bottom, as well as from the slight tapering of the walls of the fiber in the former figure.

The situation is quite different when the system is illuminated by an s-polarized electromagnetic field. In Figs. 5(a) and 5(b) we present gray level maps of the modulus of the amplitude of the single nonzero component of the electric field around the glass fiber in the absence of the metal substrate below it, and in the presence of the object depicted in Fig. 2(a) directly below it, respectively. We see in the former case, that although there is a strong electric field in the fiber itself, there is only a very weak field at the bottom of the glass fiber, which extends only a short distance into the vacuum region below its tip. This is due to the fact that for a planar waveguide consisting of a dielectric layer of thickness $d$ between perfectly conducting media there is a minimum value of the ratio $d/\lambda$, where $\lambda$ is the wavelength of the electric field in it, that has to be exceeded before an s-polarized guided wave can propagate in it. For the values of $d$ and $\lambda$ assumed in the present work we are beyond this cutoff, and almost none of the incident electromagnetic field can get through the glass fiber. Also, although this is difficult to appreciate in the figures, in the presence of the object below the fiber the electric field is somewhat stronger in the region between the tip of the fiber and the object than it is in the absence of the object.

We now turn to the images of the six structures depicted in Fig. 2. In Fig. 6 we present the near-field images of these objects obtained when the system is illuminated by a p-polarized electromagnetic field. Perhaps the most striking feature of these images is their oscillatory nature as functions of the

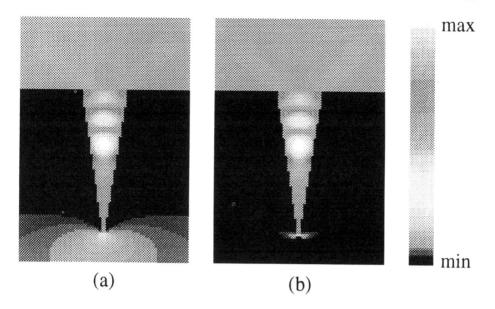

(a)          (b)

*Figure 5.* The same as Fig. 4 but for s-polarization.

coordinate $x_1$. The period of these oscillations appears to be independent of the object being imaged. Their origin will be discussed in the next section. If we seek to interpret these images in terms of the topography of the surface being scanned, perhaps the easiest one to analyze is the one plotted in Fig. 6(e), which is the image of the step depicted in Fig. 2(e). This image shows a well-defined increase as the fiber is translated above it from left to right, i.e. as the metal surface approaches the bottom of the fiber. The magnetic field impinging on the surface below the fiber is stronger the closer the surface is to the bottom of the fiber because the field decays with increasing distance from the bottom of the fiber. Thus, the field incident on the surface is stronger to the right of the step than to the left of the step, and this manifests itself in a stronger scattered intensity detected at the center of the bottom of the fiber. In the remaining images of topographic objects, i.e. in Figs. 6(a)-6(d), we note on comparing them with the objects depicted in Figs. 2(a)-2(d) that each image has a local minimum at the points where the surface profile has a maximum. This seems to be characteristic for such images. In recent calculations of similar near-field images of isolated ridges and grooves on perfectly conducting surfaces illuminated by a p-polarized, electromagnetic field[6], which calculations were carried out on the basis of the Rayleigh hypothesis[12, 13], it was found that the image of a ridge displays a peak at its position. Now, a metal characterized by a dielectric function $\epsilon(\omega) = -11.36 + i0.96$ is not a perfect conductor, but it seems to be close enough to one for the results obtained for a perfect conductor[6] to mimic those obtained here for a metal. The least informative image is the one of the dielectric discontinuity presented in Fig. 6(f). It shows that even a large increase of the magnitude of the dielectric function of the surface in crossing from the region $x_1 < 0$ into the region $x_1 > 0$, with a corresponding increase in the reflectivity of the surface, produces only a very small increase in the background intensity of the scattered field at the center of the bottom of the fiber on which the oscillations are superimposed. A dielectric step in this case is far from equivalent to a topographic step.

The images of the six objects displayed in Fig. 2 obtained under illumination by an s-polarized electromagnetic field are free from the oscillations present in the images obtained in p-polarization. These images are presented in Fig. 7, where the vertical scale has been multiplied by a factor of $10^6$ with respect to the vertical scale in Fig. 6. The features in these images correlate very well with the positions of the objects being imaged. The dips in the images plotted in Figs. 7(a)-7(d) occur at the positions where the metal surface has a maximum. The result that minima rather than maxima occur in the images at the points where the surface is closest to the fiber was also obtained in recent calculations of similar near-field images of isolated ridges and grooves on perfectly conducting surfaces by the Rayleigh method, when they are illuminated by an s-polarized electromagnetic field[6]. The image of a ridge (groove) displays a dip (peak) at its position.

Turning now to the far-field images, these have been calculated in both p- and s-polarization. However, in s-polarization the contrast in the images was comparable to the numerical precision maintained in these calculations. Consequently, we do not present them here, and display only the far-field images obtained in p-polarization. These are presented in Fig. 8. It is perhaps not surprising, in view of energy conservation considerations, that these images are complementary to the near-field images obtained in the same polarization. That is, at positions where the near-field image has a minimum the far-field image has a maximum, and *vice versa*. The information content of the far-field images thus appears to be no greater than that of the near-field images.

## 6. Conclusions

Several conclusions can be drawn from the results presented in the preceding section.

If we consider first the near field images, we see that those obtained in s-polarization correspond to lower scattered intensities and possess a smaller dynamic range than the images obtained in p-polarization. This is

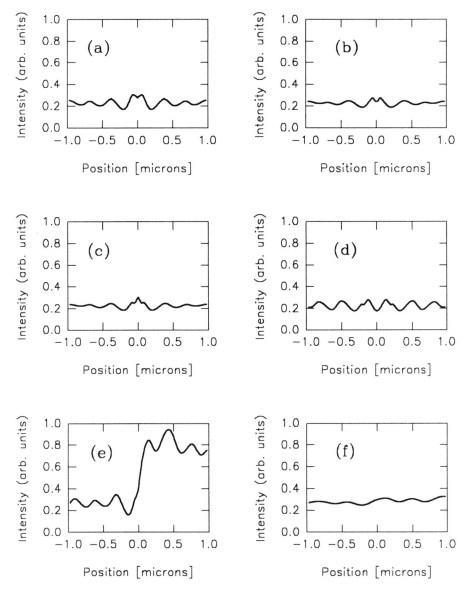

*Figure 6.* Near-field images of the six objects depicted in Fig. 2 obtained with a p-polarized electromagnetic field.

undoubtedly due to the very small fraction of the incident s-polarized electromagnetic field that is transmitted through the fiber, since we are beyond cutoff in this polarization, in comparison with the appreciable transmission of p-polarized light through it, since there is no cutoff in this polarization. However, the images obtained in s-polarization are cleaner than those ob-

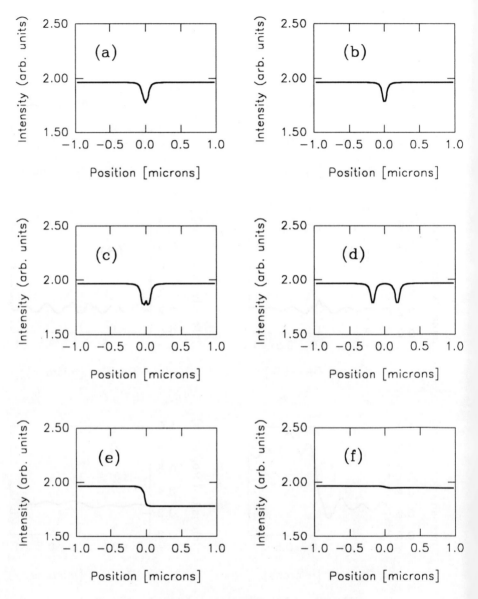

*Figure 7.* The same as Fig. 6 but for s-polarization.

tained in p-polarization, in the sense that the latter possess an oscillatory structure as functions of $x_1$ that is missing from the former. These oscillations are associated with the surface plasmon polaritons that are excited in p-polarization due to the breakdown of the translational invariance of the systems scanned caused by the presence of the object being imaged. Indeed, if we calculate the wave number $k_{sp}(\omega)$ of a surface plasmon polariton ex-

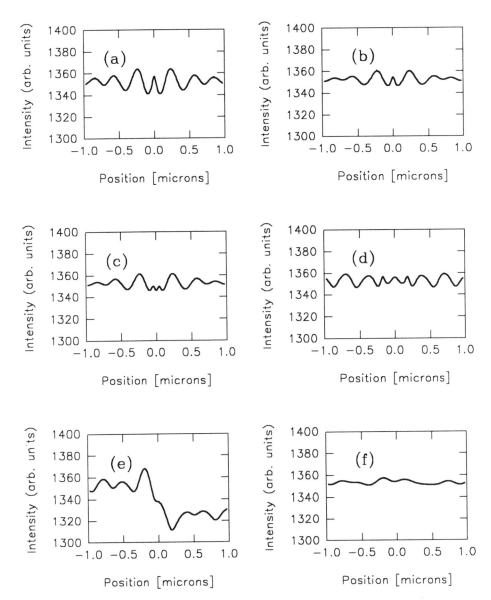

*Figure 8.* Far-field images of the six objects depicted in Fig. 2 obtained with a p-polarized electromagnetic field.

cited by an electromagnetic field of wavelength $\lambda = 0.65\mu m$ incident on a metal surface characterized by a dielectric function $\epsilon(\omega) = -11.36 + i0.96$, from the relation[14]

$$k_{sp} = \frac{\omega}{c} Re \left( \frac{\epsilon(\omega)}{\epsilon(\omega) + 1} \right)^{\frac{1}{2}}, \tag{6.1}$$

we obtain a result for the wavelength of the surface plasmon polariton along the surface of $\lambda_{sp}(\omega) = 2\pi/k_{sp}(\omega) \cong 0.62\mu m$. The period of the oscillations of the intensity (squared modulus) of the magnetic field in this surface wave as a function of $x_1$ should be half of this value, or $0.31\mu m$. This compares very well with the period of $0.32\mu m$ obtained for the oscillations observed in the results presented in Figs. 6(a) - 6(e). The absence of such oscillations in the near-field images obtained with s-polarized illumination is due to the fact that a planar metal-vacuum interface does not support surface plasmon polaritons of this polarization[14].

In addition to their relative smoothness, the near-field images obtained in s-polarization correlate better with topographic features on the surfaces being scanned by the microscope than do the images obtained in p-polarization. This is particularly evident in the images of the two rectangular ridges whose inner walls are separated by $0.05\mu m$ and $0.325\mu m$ presented in Figs. 6(c) and 6(d) in p-polarization and in Figs. 7(c) and 7(d) for s-polarization. It is difficult to find features in the images in Figs. 6(c) and 6(d) that correlate with the positions of the rectangles, for example. However, the positions of the two minima in each of Figs. 7(c) and 7(d) are close to the positions of the two rectangles in both figures. A closer look at these last two images suggests that the positions of the two minima give the positions of the centers of each of the two rectangles. The correlation between image and topography is not as clear for the other objects imaged, although the positions of the minima, and their widths, in Figs. 7(a) and 7(b) correlate well with the positions of the ridges of square and rectangular cross sections, and reasonably well with their widths.

The preceding results imply that if it is desired to obtain the lateral dimensions of a topographical feature on a metal surface from a near-field image without processing the data to obtain the surface profile itself, this can be done more readily through the use of s-polarized illumination. However, in assessing the significance of this result, one should keep in mind the comment about the somewhat artificial nature of the near-field images made at the end of Section 2.

Another conclusion that can be drawn from the results of the preceding section is that the microscope depicted in Fig. 1 can resolve topographic features on a metal surface whose lateral dimensions are of the order of $\lambda/13$, in the sense that the images in Figs. 7(c) display features (minima) that correspond to the positions of two rectangular ridges separated by that distance. Moreover, if the widths of the minima in Figs. 7(a) and 7(b) indeed represent the widths of the ridges being imaged, we can say that the resolution of our model microscope, in the sense defined above, is such that it can image features whose lateral dimensions are $\lambda/26$. It remains to be seen whether this level of resolution can be maintained when the images obtained are inverted to provide the actual profile of the surface being imaged.

In the imaging of an optical discontinuity, again the near-field image in s-polarization, Fig. 7(f), gives the position of the point at which the discontinuity of the dielectric function occurs more clearly than either the near- or far-field images obtained in p-polarization, Figs. 6(f) and 8(f), respectively. The latter images display the oscillations characteristic of the excitation of surface plasmon polaritons, which tend to obscure the point at which the optical discontinuity occurs. From Figs. 7(e) and 7(f) it is clear that, from the standpoint of the images they form in s-polarization, an optical discontinuity is equivalent to a step of reduced height on a homogeneous metal surface. This will complicate the inversion of images to obtain topographical and optical features of surfaces, unless some way can be found to discriminate between such features, perhaps on the basis of the wavelength dependence of the images, which might be different for optical features and topographic features. This point deserves further study.

Finally, we suggest the following simple physical explanations for the inversion of the profile in the near-field images obtained in both polarizations. In the case of the near-field images obtained in s-polarization the total electric field, in the structures studied here is tangent to the surface. It therefore would be required to vanish on the metal surface if the latter were in fact a perfect conductor. As we have noted above, gold at a wavelength of $0.65\mu m$ is not quite a perfect conductor, but in first approximation may be regarded as such. Thus, the scattered electric field at the top of a square or rectangular ridge just below the center of the horizontal bottom of the fiber is expected to be smaller – nearly zero – than it is at the same level in the vacuum region on either side of it. Consequently, the field scattered back into the fiber will be weaker when it is over the ridge than when it is not, leading to a dip in the image in the former case.

In the case of p-polarization it is the far-field images that seem to be easier to explain. We note that while a p-polarized plane wave incident on a planar metal surface cannot excite surface plasmon polaritons, an evanescent beam, such as the one emerging from the tip of the fiber, can. This excitation is stronger the closer the tip of the fiber is to the metal surface. It is also stronger in the vicinity of points where the surface departs sharply from planarity and loses its translational invariance, such as at the edges of a ridge or groove. The stronger the excitation of surface plasmon polaritons by the incident beam, the weaker is the scattered field in the far-field region, and *vice versa*. Thus, the far-field image of the step, presented in Fig. 8(e), is lower to the right of the step, where the fiber is closer to the metal surface, than it is to the left of the step, where it is farther from the surface, in agreement with these arguments. We note that the excitation of surface polaritons appears to be weakest when the fiber is centered at approximately $0.2\mu m$ to the left of the step, and strongest when it is at the same distance to the right of the step, as is indicated by the strong maxima and minima in the far-field image at these positions, respectively. A similar behavior is observed in the far-field images of the isolated ridges presented in Figs. 8(a) and 8(b). As the fiber approaches the ridge from the left in each case, a maximum in the far field image is observed at a distance of approximately $0.25\mu m$ from the center of the ridge, suggesting weak excitation of surface polaritons. This excitation becomes stronger as the point of discontinuity of the profile is approached, and then weakens over the flat top portion of the ridge, leading to a local maximum in the far-field intensity at the center of the ridge. As the fiber moves off to the right, the symmetry of the structure forces the far-field image to be the mirror image of what it is to the left of the ridge. This same pattern is present in the images of a pair of rectangular ridges plotted in Figs. 8(c) and 8(d), where it is seen that the images all have local maxima when the fiber is centered over the flat tops of these ridges. As we have noted earlier, maxima in far-field images imply minima in near-field images, and *vice versa*. The arguments presented here to explain the observed far-field and near-field images in p-polarization cannot be used to interpret the images obtained in s-polarization, because s-polarized surface plasmon polaritons do not exist on planar vacuum-metal interfaces[14].

## Acknowledgements

The work of AAM was supported in part by Army Research Office Grant No. DAAL 03-92-0239. AMS wishes to express his gratitude for the support of the Universidad Michoacana de San Nicolás de Hidalgo and CONACYT. MN-V thanks the Comunidad de Madrid for a travel grant to the University of California, Irvine. We would also like to express our gratitude to P. Negrete-Regagnon for his help in typesetting the figures.

# References

1. Nieto-Vesperinas, M. and Soto-Crespo, J. M. (1987) Monte Carlo Simulations for Scattering of Electromagnetic Waves From Perfectly Conductive Random Rough Surfaces, *Optics Lett.* **12**, 979-981.
2. Maradudin, A. A., Méndez, E. R., and Michel, T. (1989) Backscattering Effects in the Elastic Scattering of P-Polarized Light From a Large Amplitude Random Metallic Grating, *Optics Lett.* **14** 151-153.
3. García, N. and Nieto-Vesperinas, M. (1993) Near-Field Optics Inverse-Scattering Reconstruction of Reflective Surfaces, *Optics Lett.* **18**, 2090-2032.
4. García, N. and Nieto-Vesperinas, M. (1995) Direct Solution to the Inverse Scattering Problem for Surfaces From Near-Field Intensities Without Phase Retrieval, *Optics Lett.* **20**, 949-951.
5. Carminati, R. and Greffet, J.-J. (1995) Numerical Simulation of the Photon Scanning Tunneling Microscope, *Optics Commun.* **116**, 316-321.
6. Leskova, T. A., Maradudin A. A., and Méndez, E. R. (1996) Reconstruction of Surface Profiles From Scattering Data, paper 7 presented in Session FB1 of the National Radio Science Meeting, Boulder, Colorado, January 9-13.
7. Maradudin, A. A., Michel, T., McGurn, A. R., and Méndez, E. R. (1990) Enhanced Backscattering of Light From a Random Grating, *Ann. Phys. (N.Y.)* **203**, 255-307.
8. Danese, A. E. (1965) *Advanced Calculus, Vol. I.*, Allyn and Bacon, Inc., Boston, p. 123.
9. Mendoza-Suárez, A. and Méndez, E. R. (1995) Light Scattering by Reentrant One-Dimensional Surfaces, submitted to *J. Opt. Soc. Am. A.*
10. Garcia-Molina, R., Maradudin, A. A., and Leskova, T. A. (1990) The Impedance Boundary Condition for a Curved Surface, *Phys. Repts.* **194**, 351-359.
11. Mendoza-Suárez, A. and Méndez, E. R. (1995) An Impedance Boundary Condition for Multivalued Surface Profiles, in preparation for *Optics Commun.*
12. Rayleigh, Lord (1896) *The Theory of Sound, Vol. II, 2nd ed.* Macmillan, London, pp. 89, 96.
13. Rayleigh, Lord (1907) On the Dynamical Theory of Gratings, *Proc. Roy. Soc. London, Ser.* **A 79**, 399-416.
14. Burstein, E. Hartstein, A., Schoenwald, J., Maradudin, A. A., Mills, D. L., and Wallis, R. F. (1974) Surface Polaritons-Electromagnetic Waves at Interfaces, in E. Burstein and F. de Martini (eds.), *Polaritons*, Pergamon, New York, pp. 89-108.

# SHORT AND LONG RANGE INTERACTIONS
# IN NEAR FIELD OPTICS

O.KELLER
*Institute of Physics, Aalborg University*
*Pontoppidanstræde 103, DK-9220 Aalborg Øst, Denmark*

**Abstract.** The transition from micro to macroscopic local-field electrodynamics in optically dilute molecular media is investigated, and a new point-dipole model adequate in near field optics is established. Next, the spatial field confinement problem and its relation to near field optics are discussed taking as a starting point a new theory for the short range local-field interactions. Finally, by dividing the attached field interaction into contact and dipole-dipole interactions, the near field ($R^{-3}$-term) is identified. However, even this does not allow one to restore the classical point-dipole propagator theory.

## 1. Introduction

The present paper is written in the wake of my participation in the workshop on "Near Field Optics: Recent Progress and Perspectives", held recently in Miraflores near Madrid, and because of this it aims at describing new ideas related to the spatial field confinement problem in near field optics, the subject of my talk. However, in view of what I learned from the stimulating discussions which took place at the workshop, this article also serves the purpose of attempting to bridge (or narrow) the gap between the point-dipole model and macroscopic refractive-index approach and the microscopic local-field theory.

In section 2.1, the microscopic theory of local-field electrodynamics in mesoscopic particles and molecules (atoms) is briefly reviewed in a manner which emphasizes the physical content, and in section 2.2, the theory is applied to a study of a condensed system of optically active but electronically decoupled molecules. By dividing the electrodynamic interactions

63

*M. Nieto-Vesperinas and N. García (eds.), Optics at the Nanometer Scale 63–93.*
© *1996 Kluwer Academic Publishers. Printed in the Netherlands.*

into short and long range ones, a new point-dipole model is established for so-called optically dilute molecular systems, i.e. media in which optically induced short range interactions between neighbouring molecules play no role. The relation of the rigorously established new point-dipole model to the standard ones used previously in near field optics is discussed, and the transition to the phenomenological macroscopic approaches investigated. In the new point-dipole theory, short range propagation effects hidden in the $R^{-3}$-term, and to some extent in the $R^{-2}$-term also, are included in the dressed molecular polarizability. In sections 3.1 and 3.2, the spatial field confinement problem and its relation to near field optics are analysed. By paying attention to the spatial structure of the so-called attached fields, it is shown that by leaving out the contact interactions and keeping only the tail of the dipole-dipole interaction one may insist in using a $R^{-3}$-term in the new propagator. Readers interested in the application of the local-field theory to near field phaseconjugation, a subject also discussed in my Miraflores talk, is referred to my upcoming paper from the "2nd Mediterranean Workshop and Topical Meeting on Novel Optical Materials and Applications" held in Cetraro in Italy this year. The paper which is entitled "Quantum Dots of Light" deals with the theory for phaseconjugation of near fields, so as to form light dots (foci) of subwavelength extension, and will be published in a special issue (volume 5, Nos 1 and 2) of J.Nonlinear Optical Physics and Materials (World Scientific, 1996).

## 2. From micro to macroscopic local-field electrodynamics

Up to now the overwhelming majority of the theoretical studies in near field optics has aimed at carrying out selfconsistent field calculations on the basis of either the macroscopic refractive index concept, or heuristic microscopic models treating matter as an assembly of point particles interacting through multipole fields. Each of the two afore-mentioned approaches has it own merits and demerits, and both have contributed in an essential manner to improve our understanding of the role played by local fields in near field optics [1].

Near field optical experiments have demonstrated that it is possible to achieve a spatial resolution substantially better than predicted by classical diffraction theory, and it seems to me that the most important goal ahead of us is an investigation of the possibilities for approaching (reaching) atomic resolution. Even from a conceptual point of view it is an open question whether or not a spatial resolution on the atomic length scale is conceivable after all. To adress this question theoretically it is necessary to go beyond the frameworks offered by refractive-index approaches and point-dipole models. Microscopic local-field theories developed in recent years to describe the electrodymamics of mesoscopic media appear adequate for investigations of the spatial field confinement problem on the atomic scale, and for finding in actual cases the length-scale limit beyond which macroscopic theories fail. In passing let me stress that in my opinion there is no universal length scale separating the micro and macroscopic domains.

In the present section I shall briefly review in physical terms the microscopic local-field approach in a manner adequate in the context of near field optics, and then demonstrate how the frameworks for the existing macroscopic theories can be derived rigorously. For a detailed account of local-field electrodynamics in mesoscopic media the reader is referred to my upcoming article in Physics Reports [2].

## 2.1. THE MICROSCOPIC THEORY

Let us assume that the system of material particles under study is exposed to a prescribed monochromatic external (ext) electromagnetic field, $\vec{E}^{ext}(\vec{r};\omega)$, of cyclic frequency $\omega$. Described in the frequency domain, the external field gives rise to a local electric field, $\vec{E}(\vec{r};\omega)$, given by [2,3]

$$\vec{E}(\vec{r};\omega) = \vec{E}^{ext}(\vec{r};\omega) - i\mu_0\omega \int \overset{\leftrightarrow}{G_0}(\vec{r} - \vec{r}';\omega) \cdot \vec{J}(\vec{r}';\omega)d^3r' , \qquad (1)$$

where $\overset{\leftrightarrow}{G_0}(\vec{r} - \vec{r}';\omega)$ is the electromagnetic propagator, consisting of the sum of the vacuum propagator and the self-field terms, and $\vec{J}(\vec{r}';\omega)$ is the field-induced microscopic current density prevailing at the point $\vec{r}'$ in our system. In physical terms, the interpretation of Eq.(1) is straightforward. Thus, to obtain the local field at the observation point $\vec{r}$, one just adds to the external field at $\vec{r}$, the fields created by the various induced current oscillations ( $\partial \vec{J}(\vec{r}',t')/\partial t')d^3r' \Longrightarrow -i\omega\vec{J}(\vec{r}';\omega)d^3r'$ prevailing in the infinitesimal volume element $d^3r'$ located at the source point $\vec{r}'$. In the frequency domain the field propagation between the source ($\vec{r}'$) and observation ($\vec{r}$) points is governed by the electromagnetic Green's function $\overset{\leftrightarrow}{G_0}(\vec{r} - \vec{r}';\omega)$. If the field-induced current density distribution were known the local field could readily be obtained from Eq.(1) by a direct integration, and vice versa, knowing $\vec{E}(\vec{r}';\omega)$ would allow one to calculate $\vec{J}(\vec{r};\omega)$ from

$$\vec{J}(\vec{r};\omega) = \frac{i}{\mu_0\omega} \int \overset{\leftrightarrow}{G_0}{}^{-1}(\vec{r} - \vec{r}';\omega) \cdot \left[\vec{E}(\vec{r}';\omega) - \vec{E}^{ext}(\vec{r}';\omega)\right] d^3r' , \qquad (2)$$

where the so-called inverse electromagnetic propagator $\overset{\leftrightarrow}{G_0}{}^{-1}(\vec{r} - \vec{r}';\omega)$ is defined via the integral equation

$$\int \overset{\leftrightarrow}{G_0^{-1}} (\vec{r} - \vec{r}''; \omega) \cdot \overset{\leftrightarrow}{G_0}(\vec{r}'' - \vec{r}'; \omega) d^3 r'' = \overset{\leftrightarrow}{U} \delta(\vec{r} - \vec{r}') , \qquad (3)$$

$\overset{\leftrightarrow}{U}$ being the unit tensor (of dimension 3 × 3), and $\delta(\vec{r} - \vec{r}')$ the Dirac delta function. As expected the induced current density is related to the *induced* field, $\vec{E} - \vec{E}^{ext}$, only.

Facing the fact that the microscopic Maxwell-Lorentz equations, leading to Eq.(1), end up giving an integral relation between two unknown distributions, viz. $\vec{E}(\vec{r}; \omega)$ and $\vec{J}(\vec{r}; \omega)$, an extra relation between $\vec{E}$ and $\vec{J}$ is needed to break the vicious circle. In nonrelativistic (quantum) electrodynamics the Schrödinger equation provides us with such a relation. Within the framework of linear response theory, to which we shall limit ourselves here, one finds [2]

$$\vec{J}(\vec{r}; \omega) = \int \overset{\leftrightarrow}{\sigma}_{MB} (\vec{r}, \vec{r}'; \omega) \cdot \left[ \vec{E}_T(\vec{r}'; \omega) + \vec{E}_L^{ext}(\vec{r}'; \omega) \right] d^3 r' , \qquad (4)$$

where $\overset{\leftrightarrow}{\sigma}_{MB}(\vec{r}, \vec{r}'; \omega)$ is the linear and nonlocal many-body conductivity tensor of the particle system, $\vec{E}_T(\vec{r}'; \omega)$ is the divergence-free [transverse(T)] part of the local field, and $\vec{E}_L^{ext}(\vec{r}'; \omega)$ is the rotational-free [longitudinal (L)] part of the external field. In cases where the external sources are located outside the so-called longitudinal (or equivalently transverse) current-density domain, Eq.(4) is reduced to the form [2]

$$\vec{J}(\vec{r}; \omega) = \int \overset{\leftrightarrow}{\sigma}_{MB} (\vec{r}, \vec{r}'; \omega) \cdot \vec{E}_T(\vec{r}'; \omega) d^3 r' . \qquad (5)$$

The reason that only the transverse part of the local field enters the constitutive relation is associated with the fact that the redundant dynamical part of the scalar potential (and here the longitudinal part of the local field) can be eliminated in favour of the particle position variables, using the Coulomb gauge [4]. By combining Eqs.(1) and (4) [or if adequate Eq.(5)], a loop equation can be established for the transverse part of the local field. Once $\vec{E}_T$ has been determined from this equation, the longitudinal part of the local field, $\vec{E}_L(\vec{r}; \omega)$, can be obtained by direct integration [2].

Having in mind here the transition from microscopic to macroscopic electrodynamics I shall not dwell here on the many-body local-field electrodynamics. Interested readers may be referred to Ref.[2]. In practice it is difficult to carry out a rigorous calculation of the many-body conductivity because the many-body wave functions entering $\overset{\leftrightarrow}{\sigma}_{MB}(\vec{r},\vec{r}';\omega)$ are hard (or impossible) to determine for systems having more than a few mobile particles. In the random-phase-approximation (RPA) approach, which is sufficient for the present purpose, one assumes that the induced current-density response of the electrons effectively is to the total local field $\vec{E}(\vec{r};\omega) = \vec{E}_T(\vec{r};\omega) + \vec{E}_L(\vec{r};\omega)$ instead of to the field combination $\vec{E}_T(\vec{r};\omega) + \vec{E}_L^{ext}(\vec{r};\omega)$, and that the corresponding conductivity $\overset{\leftrightarrow}{\sigma}_{RPA}(\vec{r},\vec{r}';\omega) \equiv \overset{\leftrightarrow}{\sigma}(\vec{r},\vec{r}';\omega)$ is a one-particle response tensor [5]. In the RPA theory, the constitutive relation thus takes the form

$$\vec{J}(\vec{r};\omega) = \int \overset{\leftrightarrow}{\sigma}(\vec{r},\vec{r}';\omega) \cdot \vec{E}(\vec{r}';\omega) d^3r' . \tag{6}$$

By inserting the expression for $\vec{J}(\vec{r};\omega)$ given in the equation above into Eq.(1) one obtains the following integral equation for the local field:

$$\vec{E}(\vec{r}) = \vec{E}^{ext}(\vec{r}) - i\mu_0\omega \int \overset{\leftrightarrow}{G}_0(\vec{r} - \vec{r}') \cdot \overset{\leftrightarrow}{\sigma}(\vec{r}',\vec{r}'') \cdot \vec{E}(\vec{r}'') d^3r'' d^3r' , \tag{7}$$

omitting for notational simplicity the reference to $\omega$ from the notation.

To proceed from here an explicit expression for the conductivity tensor is needed. Neglecting the spin part of $\overset{\leftrightarrow}{\sigma}(\vec{r},\vec{r}';\omega)$ one may take

$$\overset{\leftrightarrow}{\sigma}(\vec{r},\vec{r}';\omega) = \frac{2i}{\omega} \sum_{m,n} \left( \frac{\hbar\omega}{\varepsilon_n - \varepsilon_m} \right) \frac{f_m - f_n}{\hbar\omega + \varepsilon_m - \varepsilon_n} \vec{j}_{nm}(\vec{r})\vec{j}_{mn}(\vec{r}') \tag{8}$$

in dyadic notation. The quantities $\vec{j}_{nm}(\vec{r})$ and $\vec{j}_{mn}(\vec{r}')$ are the transition current densities between the one-electron energy eigenstates $m$ and $n$, calculated at the space points $\vec{r}$ and $\vec{r}'$, respectively. The explicit expression for the transition current density *from* state $\alpha$ *to* state $\beta$ is given by

$$\vec{j}_{\alpha\beta}(\vec{r}) = \frac{e\hbar}{2mi} \left( \psi_\alpha(\vec{r})\vec{\nabla}\psi_\beta^*(\vec{r}) - \psi_\beta^*(\vec{r})\vec{\nabla}\psi_\alpha(\vec{r}) \right) , \tag{9}$$

where $\psi_\alpha$ and $\psi_\beta$ are the wave functions belonging to the energy eigenstates, the respective energies being denoted by $\varepsilon_\alpha$ and $\varepsilon_\beta$. The quantities $f_m$ and $f_n$ are Fermi-Dirac distribution factors giving the probabilities that the states $m$ and $n$ are occupied in thermal equilibrium. Physically, the transition current density in Eq.(9) has the following meaning. Starting from the standard expression for the quantum mechanical current density, i.e.

$$\vec{j}(\vec{r},t) = \frac{e\hbar}{2im} \left( \psi(\vec{r},t)\vec{\nabla}\psi^*(\vec{r},t) - \psi^*(\vec{r},t)\vec{\nabla}\psi(\vec{r},t) \right) , \qquad (10)$$

and inserting a *pure state* wave function which is a linear combination of $\psi_\alpha$ and $\psi_\beta$, viz.

$$\psi(\vec{r},t) = c_\alpha(t)\psi_\alpha(\vec{r}) + c_\beta(t)\psi_\beta(\vec{r}) ,$$

with $| c_\alpha |^2 + | c_\beta |^2 = 1$ (normalization), the current density can be written in the form

$$\vec{j}(\vec{r},t) = \vec{j}_{\alpha\alpha}(\vec{r}) | c_\alpha(t) |^2 + \vec{j}_{\beta\beta}(\vec{r}) | c_\beta(t) |^2 + \vec{j}_{\alpha\beta}(\vec{r})c_\alpha(t)c_\beta^*(t)$$

$$+ \vec{j}_{\beta\alpha}(\vec{r})c_\alpha^*(t)c_\beta(t) . \qquad (11)$$

It appears from Eq.(11) that $\vec{j}_{\alpha\beta}(\vec{r})$ is directly related to the current density flow accompanying an electronic transition *from* state $\alpha$ *to* state $\beta$. Roughly speaking $\vec{j}_{\alpha\beta}(\vec{r})$ represents the *spatial* part of this flow, and the product $c_\alpha(t)c_\beta^*(t)$ the dynamic time evolution of the flow. In the same manner, the term $\vec{j}_{\beta\alpha}(\vec{r})c_\alpha^*(t)c_\beta(t)$ is associated with the current density flow accompanying the $\beta$ *to* $\alpha$ transition. Finally, the $\vec{j}_{\alpha\alpha} | c_\alpha(t) |^2$ and $\vec{j}_{\beta\beta} | c_\beta(t) |^2$ terms describe the time development of the current density flows in the $\alpha$ and $\beta$ eigenstates. In many cases, $\vec{j}_{\alpha\alpha} = \vec{j}_{\beta\beta} = \vec{0}$.

It is worthwhile here to make two further comments on the expression for $\overleftrightarrow{\sigma} (\vec{r}, \vec{r}'; \omega)$. Firstly, the expression contains both the diamagnetic and paramagnetic contributions to the linear response, see e.g. Ref.[6]. Thus, if the factor $\hbar\omega/(\varepsilon_n - \varepsilon_m)$ is replaced by unity, Eq.(8) gives the paramagnetic part of the conductivity tensor, only. Secondly, one may take electronic relaxation processes into account phenomenologically by replacing the denominator $\hbar\omega + \varepsilon_m - \varepsilon_n$ by $\hbar(\omega + i\gamma_{mn}) + \varepsilon_m - \varepsilon_n$, where $\gamma_{mn}$ is the appropriate damping constant for transition between the $m$ and $n$ states.

Let us now return to the integral equation for the local field. Hence, by inserting Eq.(8) into Eq.(7) one obtains

$$\vec{E}(\vec{r}) = \vec{E}^{ext}(\vec{r}) + 2\mu_0 \sum_{m,n}(\frac{\hbar\omega}{\varepsilon_n - \varepsilon_m})\frac{f_m - f_n}{\hbar\omega + \varepsilon_m - \varepsilon_n}\beta_{mn}$$

$$\times \int \overleftrightarrow{G}_0(\vec{r} - \vec{r}') \cdot \vec{j}_{nm}(\vec{r}')d^3r' , \qquad (12)$$

where

$$\beta_{mn} = \int \vec{j}_{mn}(\vec{r}'') \cdot \vec{E}(\vec{r}'') d^3 r'' \; . \tag{13}$$

For conceptual purposes it is convenient at this stage to rewrite Eq.(12) as follows:

$$\vec{E}(\vec{r}) = \vec{E}^{\,ext}(\vec{r}) - i\mu_0\omega \sum_{m,n} \int \overleftrightarrow{G}_0(\vec{r} - \vec{r}') \cdot \vec{I}_{nm}(\vec{r}') d^3 r' \; , \tag{14}$$

with

$$\vec{I}_{nm}(\vec{r}') = \frac{2i\hbar}{\varepsilon_n - \varepsilon_m} \frac{f_m - f_n}{\hbar\omega + \varepsilon_m - \varepsilon_n} \beta_{mn} \vec{j}_{nm}(\vec{r}') \; . \tag{15}$$

At this point a simple physical interpretation of Eq.(14) emerges. To realize this let us focus our attention on a particular electronic transition say *from n to m*. Associated with this transition one has a current density distribution $\vec{I}_{nm}(\vec{r}')$ which radiates. To obtain the contribution to the local field at $\vec{r}$ from $\vec{I}_{nm}(\vec{r}')$ one just calculates the integral $-i\mu_0\omega \int \overleftrightarrow{G}_0(\vec{r} - \vec{r}')$ $\cdot \vec{I}_{nm}(\vec{r}') d^3 r'$, cf. Eq.(1). To determine to the local field, $\vec{E}(\vec{r})$, at $\vec{r}$, one adds to the external field, $\vec{E}^{\,ext}(\vec{r})$, at $\vec{r}$, the induced field radiation from all (summation over $n$ and $m$) relevant electronic transitions, each transition being accompanied by a current density distribution $\vec{I}_{nm}(\vec{r}')$. Provided the various $\vec{I}_{nm}(\vec{r}')$'s were known the local field thus could be obtained directly from Eq.(14). To carry out the spatial integration $\int \overleftrightarrow{G}_0 (\vec{r} - \vec{r}')$ $\cdot \vec{I}_{nm}(\vec{r}') d^3 r'$ one just needs to know the spatial form of $\vec{I}_{nm}(\vec{r}')$. It appears from Eq.(15) that this form is precisely the form of the transition current density $\vec{j}_{nm}(\vec{r}')$, which in fact is known. So to determine the local field, at this stage one only needs to determine the various $\beta_{mn}$ numbers.

Essentially, the $\beta_{mn}$'s thus give the amplitude strengths of the radiation from the various $n$ *to* $m$ transitions. If we focus our attention on the radiation from a particular transition, say the $n$ *to* $m$ transition, it appears from Eq.(13) that the related amplitude strength $\beta_{nm}$ is proportional to the work performed by the field in order to excite the system from state $m$ *to* state $n$. The spatial distribution of the current density flow accompanying the $m$ *to* $n$ transition is given by the transition current density $\vec{j}_{mn}(\vec{r}')$, and the field which acts upon this flow is the yet unknown local field, $\vec{E}(\vec{r}')$. This is so because the field carrying out the work necessarily must be the sum of the external field and the fields radiated by all other

70

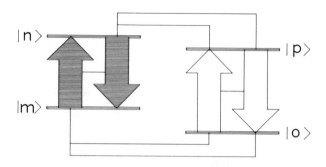

*Figure 1.* Schematic diagram illustrating the local-field coupling between two sets
($| m > \rightleftharpoons | n >$, $| o > \rightleftharpoons | p >$ ) of electronic transitions. The thin lines indicate that the
$| m > \rightarrow | n >$ absorption process as well as the $| n > \rightarrow | m >$ emission process are cou-
pled to both the $| o > \rightarrow | p >$ (absorption) and $| p > \rightarrow | o >$ (emission) transition. Thin
lines also indicate the presence of electrodynamic self-coupling in the $| m > \rightleftharpoons | n >$ and
$| o > \rightleftharpoons | p >$ transitions.

relevant electronic transitions, cf. Fig.1. Since the absorption and emission
processes accompanying the various electronic transitions hence are mutually coupled, a selfconsistent determination of the various $\beta_{mn}$'s is needed.
Before calculating the unknown field strengths $\beta_{mn}$, let us rewrite Eq.(14)
in the form

$$\vec{E}(\vec{r}) = \vec{E}^{ext}(\vec{r}) + \sum_{m,n} \beta_{mn} \vec{F}_{nm}(\vec{r}) \ , \tag{16}$$

where

$$\vec{F}_{nm}(\vec{r}) = 2\mu_0 \frac{\hbar\omega}{\varepsilon_n - \varepsilon_m} \frac{f_m - f_n}{\hbar\omega + \varepsilon_m - \varepsilon_n} \int \overleftrightarrow{G}_0(\vec{r} - \vec{r}') \cdot \vec{j}_{nm}(\vec{r}')d^3r' \tag{17}$$

in principle is a known function of $\vec{r}$. By substituting the ansatz in Eq.(16)
back into Eq.(13) one obtains

$$\beta_{mn} - \sum_{o,p} N_{op}^{mn} \beta_{op} = H_{mn} \ , \tag{18}$$

where we have introduced the quantities

$$N_{op}^{mn} = \int \vec{j}_{mn}(\vec{r}) \cdot \vec{F}_{po}(\vec{r})d^3r \ , \tag{19}$$

and

$$H_{mn} = \int \vec{j}_{mn}(\vec{r}) \cdot \vec{E}^{\,ext}(\vec{r})d^3r \ . \tag{20}$$

By letting $(m,n)$ run through all possible (relevant) combinations, Eq.(18) gives a set of inhomogeneous algebraic equations among the unknown $\beta_{mn}$'s. By taking into account a finite number of levels, viz. the most relevant ones in a given situation, the determination of the local field from the integral equation in (7) has been reduced to solving a matrix equation problem for the unknown field strengths, $\beta_{mn}$. In passing it is worth noticing that the discretization of the integral equation problem for the local field in the microscopic theory is done on the basis of the electronic level (eigenstate) scheme whereas it in macroscopic approaches is made on the basis of small volume elements in direct space, see Ref.[1] and section 2.2.5.

In the absence of the external field, all $H_{mn}$'s are zero. The solutions to the resulting homogeneous set of equations in (18) represent the possible local-field resonances of the system in consideration. Thus, from the condition that the determinant (Det $\{...\}$ ) of the system of homogeneous equations vanishes, i.e.

$$Det\left\{\delta_{mn,op} - N_{op}^{mn}(\omega^{RES})\right\} = 0 \ , \tag{21}$$

where $\delta_{mn,op}$ is the Krönecker delta, the local-field resonance frequencies $\omega^{RES}$ may be obtained [2].

## 2.2. THE MOLECULAR LIMIT

### 2.2.1. *Electronic decoupling*
Let us now consider the case where our system is composed of electronically uncoupled molecules. By numbering the individual molecules with the index $j$, the conductivity tensor takes the form

$$\overleftrightarrow{\sigma}(\vec{r},\vec{r}\,') = \sum_j \overleftrightarrow{\sigma}^j(\vec{r},\vec{r}\,') \ , \tag{22}$$

where

$$\overleftrightarrow{\sigma}^j(\vec{r},\vec{r}\,') = \frac{2i}{\omega} \sum_{m_j,n_j} \left(\frac{\hbar\omega}{\varepsilon_{n_j} - \varepsilon_{m_j}}\right) \frac{f_{m_j} - f_{n_j}}{\hbar\omega + \varepsilon_{m_j} - \varepsilon_{n_j}} \vec{j}_{n_j m_j}(\vec{r})\vec{j}_{m_j n_j}(\vec{r}\,')$$

$$\tag{23}$$

is the response function of molecule number $j$, the energy eigenstates of this molecule being classified by the quantum numbers $m_j$ and $n_j$. The electronic decoupling of the various molecules implies that the transition current densities $\vec{\jmath}_{n_j m_j}(\vec{r})$ and $\vec{\jmath}_{m_j n_j}(\vec{r}')$ are different from zero only when the points $\vec{r}$ and $\vec{r}'$ are located inside molecule number $j$.

Following the procedure described in section 2.1, it appears that the local field for electronically decoupled molecular system is given by

$$\vec{E}(\vec{r}) = \vec{E}^{\,ext}(\vec{r}) + \sum_j \sum_{m_j,n_j} \beta_{m_j n_j} \vec{F}_{n_j m_j}(\vec{r}) \;, \tag{24}$$

where

$$\vec{F}_{n_j m_j}(\vec{r}) =$$

$$2\mu_0 \frac{\hbar\omega}{\varepsilon_{n_j} - \varepsilon_{m_j}} \frac{f_{m_j} - f_{n_j}}{\hbar\omega + \varepsilon_{m_j} - \varepsilon_{n_j}} \int \overleftrightarrow{G}_0(\vec{r} - \vec{r}') \cdot \vec{\jmath}_{n_j m_j}(\vec{r}') d^3 r' \;, \tag{25}$$

and

$$\beta_{m_j n_j} = \int \vec{\jmath}_{m_j n_j}(\vec{r}) \cdot \vec{E}(\vec{r}) d^3 r \;. \tag{26}$$

Although we have made an electronic decoupling of the individual molecules this does not imply that the field strength $\beta_{m_i n_i}$ with which molecule number $i$ radiates is independent of the presence of the remaining molecules, identified by the running index $j$. In fact, the various field strengths satisfy the set of algebraic equations

$$\beta_{m_i n_i} - \sum_{o_j, p_j} N^{m_i n_i}_{o_j p_j} \beta_{o_j, p_j} = H_{m_i n_i} \;, \tag{27}$$

where

$$N^{m_i n_i}_{o_j p_j} = \int \vec{\jmath}_{m_i n_i}(\vec{r}) \cdot \vec{F}_{p_j, o_j}(\vec{r}) d^3 r \tag{28}$$

and

$$H_{m_i n_i} = \int \vec{\jmath}_{m_i n_i}(\vec{r}) \cdot \vec{E}^{\,ext}(\vec{r}) d^3 r \;, \tag{29}$$

cf. Eqs.(18)-(20). Terms in Eq.(28) for which $i = j$ represent the electrodynamic self-coupling within the individual molecules, and terms with $i \neq j$

are associated with the mutual electromagnetic interaction between different molecules. It appears from Eq.(27) that the field strengths in general are
determined by both the self-coupling and the mutual molecular interaction.

### 2.2.2. Short and long range electrodynamic interactions

To gain deeper insight into the physics determining the local field let us
consider the structure of the electromagnetic propagator, $\overleftrightarrow{G}_0(\vec{r} - \vec{r}';\omega)$.
From the analysis of e.g. Ref.[2] it appears that

$$\overleftrightarrow{G}_0(\vec{r} - \vec{r}') = \overleftrightarrow{D}_0^T(\vec{r} - \vec{r}') + \overleftrightarrow{g}^L(\vec{r} - \vec{r}') + \overleftrightarrow{g}^T(\vec{r} - \vec{r}') , \qquad (30)$$

where

$$\overleftrightarrow{g}^T(\vec{r} - \vec{r}') = \frac{1}{3}\left(\frac{c_0}{\omega}\right)^2 \overleftrightarrow{\delta}^T(\vec{r} - \vec{r}') , \qquad (31)$$

and

$$\overleftrightarrow{g}^L(\vec{r} - \vec{r}') = \left(\frac{c_0}{\omega}\right)^2 \overleftrightarrow{\delta}^L(\vec{r} - \vec{r}') , \qquad (32)$$

are the so-called transverse and longitudinal self-field propagators, respec-
tively, $\overleftrightarrow{\delta}^T$ and $\overleftrightarrow{\delta}^L$ being the transverse and longitudinal Dirac delta functions (tensors), and [2]

$$\overleftrightarrow{D}_0^T(\vec{r} - \vec{r}') -$$

$$\frac{q_0}{4\pi i}\left(\left\{\frac{1}{iq_0 R}\left(\overleftrightarrow{U} - \vec{e}_{\vec{R}}\vec{e}_{\vec{R}}\right) - \left[\frac{1}{(iq_0 R)^2} - \frac{1}{(iq_0 R)^3}\right]\left(\overleftrightarrow{U} - 3\vec{e}_{\vec{R}}\vec{e}_{\vec{R}}\right)\right\} e^{iq_0 R}\right.$$

$$\left. - \frac{1}{(iq_0 R)^3}\left(\overleftrightarrow{U} - 3\vec{e}_{\vec{R}}\vec{e}_{\vec{R}}\right)\right) \qquad (33)$$

is the transverse part of the vacuum propagator. In Eq.(33), the abbrevia-
tions $q_0 = \omega/c_0$, $\vec{r} - \vec{r}' = \vec{R}$, and $\vec{e}_{\vec{R}} = \vec{R}/R$ are used. Since [2]

$$\overleftrightarrow{g}^L(\vec{R}) = -3\,\overleftrightarrow{g}^T(\vec{R}) = \frac{q_0}{4\pi i}\frac{1}{(iq_0 R)^3}\left(\overleftrightarrow{U} - 3\vec{e}_{\vec{R}}\vec{e}_{\vec{R}}\right) , \qquad R \neq 0 , \qquad (34)$$

it is seen that the self-field propagators carry short range electrodynamic
interactions only, and that these are of the electric dipole-dipole (near field)
interaction type. The long range electrodynamic interactions are carried by
the transverse vacuum propagator, which in the far field takes the usual
form

$$\overset{\leftrightarrow T}{D_0}(\vec{R}) = -\frac{e^{iq_0 R}}{4\pi R}\left(\overset{\leftrightarrow}{U} - \vec{e}_{\vec{R}}\vec{e}_{\vec{R}}\right) , \qquad q_0 R \gg 1 . \tag{35}$$

Since

$$\overset{\leftrightarrow T}{D_0}(\vec{R}) = \frac{q_0}{4\pi i}\left[\frac{i}{2\pi q_0 R}\left(\overset{\leftrightarrow}{U} + \vec{e}_{\vec{R}}\vec{e}_{\vec{R}}\right) + \frac{2}{3}\overset{\leftrightarrow}{U} + O(q_0 R)\right] , \tag{36}$$

in the limit $q_0 R \to 0$, it is realized that the near field electrodynamics is dominated by the self-field parts of the propagator. In the context of retardation, the self-field contributions to $\overset{\leftrightarrow}{G}_0$ are seen to provide nonretarded electrodynamic interactions (of the order $c_0^2$), whereas $\overset{\leftrightarrow T}{D_0}$ of Eq.(36) gives an in-phase retarded second-order correction (of the order $c_0^2$) and an out-of-phase retarded third-order correction (of the order $c_0^{-1}$).

In view of the afore-mentioned circumstances it is adequate to divide the $\vec{F}_{n_j m_j}$'s of Eq.(25) into short (SR) and long range (LR) parts, i.e.

$$\vec{F}_{n_j m_j}(\vec{r}) = \vec{F}^{SR}_{n_j m_j}(\vec{r}) + \vec{F}^{LR}_{n_j m_j}(\vec{r}) , \tag{37}$$

where

$$\vec{F}^{SR}_{n_j m_j}(\vec{r}) =$$

$$\frac{2}{\varepsilon_0 \omega^2}\frac{\hbar\omega}{\varepsilon_{n_j} - \varepsilon_{m_j}}\frac{f_{m_j} - f_{n_j}}{\hbar\omega + \varepsilon_{m_j} - \varepsilon_{n_j}}\left(\frac{1}{3}\vec{j}^{T}_{n_j m_j}(\vec{r}) + \vec{j}^{L}_{n_j m_j}(\vec{r})\right) , \tag{38}$$

and

$$\vec{F}^{LR}_{n_j m_j}(\vec{r}) =$$

$$2\mu_0 \frac{\hbar\omega}{\varepsilon_{n_j} - \varepsilon_{m_j}}\frac{f_{m_j} - f_{n_j}}{\hbar\omega + \varepsilon_{m_j} - \varepsilon_{n_j}}\int \overset{\leftrightarrow T}{D_0}(\vec{r} - \vec{r}') \cdot \vec{j}_{n_j m_j}(\vec{r}')d^3 r' . \tag{39}$$

In Eq.(38), $\vec{j}^{T}_{n_j m_j}(\vec{r})$ and $\vec{j}^{L}_{n_j m_j}(\vec{r})$ denote the divergence-free and rotational-free parts of the transition current density $\vec{j}_{n_j m_j}(\vec{r})$, respectively. By inserting Eq.(37) into Eq.(24), the result for the local field can be written in the form

$$\vec{E}(\vec{r}) = \vec{E}^{ext}(\vec{r}) + \sum_j \sum_{m_j n_j} \beta_{m_j n_j} \vec{F}^{SR}_{n_j m_j}(\vec{r}) + \sum_j \sum_{m_j n_j} \beta_{m_j n_j} \vec{F}^{LR}_{n_j m_j}(\vec{r}) . \tag{40}$$

Although the $\vec{F}_{n_j m_j}$'s may be split into short and long range terms, the amplitude strengths $\beta_{m_j n_j}$ still have to be determined from a set of algebraic

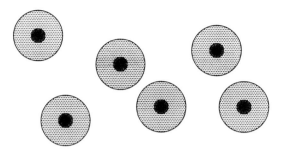

*Figure 2.* Schematic illustration of an optically dilute molecular medium. The domains occupied by the (relevant) transition current densities of the individual molecules are represented by the black areas, and the ranges of the nonoverlaping short range fields of the individual molecule are indicated by the grey domains.

equations where the individual matrix elements $N_{o_j p_j}^{m_i n_i}$ contain additive contributions from SR and LR interactions, viz.

$$N_{o_j p_j}^{m_i n_i}(SR) = \frac{2}{\varepsilon_0 \omega^2} \frac{\hbar \omega}{\varepsilon_{n_i} - \varepsilon_{m_i}} \frac{f_{m_i} - f_{n_i}}{\hbar \omega + \varepsilon_{m_i} - \varepsilon_{n_i}}$$
$$\times \int \vec{j}_{m_i n_i}(\vec{r}) \cdot \left( \frac{1}{3} \vec{j}_{p_j o_j}^T(\vec{r}) + \vec{j}_{p_j o_j}^L(\vec{r}) \right) d^3 r \ , \quad (41)$$

and

$$N_{o_j p_j}^{m_i n_i}(LR) = 2\mu_0 \frac{\hbar \omega}{\varepsilon_{n_i} - \varepsilon_{m_i}} \frac{f_{m_i} - f_{n_i}}{\hbar \omega + \varepsilon_{m_i} - \varepsilon_{n_i}}$$
$$\times \int \vec{j}_{m_i n_i}(\vec{r}) \cdot \overset{\leftrightarrow}{D}_0^T(\vec{r} - \vec{r}') \cdot \vec{j}_{p_j o_j}(\vec{r}') d^3 r' d^3 r \ . \quad (42)$$

### 2.2.3. *Optically dilute molecular systems*

Faced with the fact that it in general is an impossible task, even in the molecular limit, to solve directly the set of algebraic equations in (27), it is of interest to investigate particularly simple cases to gain further insight into the physics behind local-field theories. Let us hence consider a particle system in which the optically active molecules are so far apart that the short range interaction between *different* molecules can be neglected, see Fig.2. I name such a particle arrangement an optically dilute molecular medium.

As a starting point, Eqs.(22) and (30) are inserted into Eq.(7). This leads to the following equation for the local field:

$$\vec{E}(\vec{r}) =$$

$$\vec{E}^{\,ext}(\vec{r}) - i\mu_0\omega \sum_j \int \overset{\leftrightarrow T}{D_0}(\vec{r} - \vec{r}\,') \cdot \overset{\leftrightarrow j}{\sigma}(\vec{r}\,', \vec{r}\,'') \cdot \vec{E}(\vec{r}\,'')d^3r''d^3r'$$

$$-i\mu_0\omega \sum_j \int \left[\overset{\leftrightarrow T}{g}(\vec{r} - \vec{r}\,') + \overset{\leftrightarrow L}{g}(\vec{r} - \vec{r}\,')\right] \cdot \overset{\leftrightarrow i}{\sigma}(\vec{r}\,', \vec{r}\,'') \cdot \vec{E}(\vec{r}\,'')d^3r''d^3r' \ .$$

$$(43)$$

Next, we focuse our attention on the local field *inside* a particular molecule, say number $i$, and we rename the corresponding observation coordinate $\vec{r}_i$. Since we are dealing with an optically dilute molecular medium, among the short range terms in Eq.(43) only the term coming from the molecule itself (number $i$) survives. Furthermore, due to the fact that the SR contribution inside the molecule in consideration is much large than the LR contribution from the same molecule, one may safely neglect the $j = i$ term among the long range terms of Eq.(43). As we shall realize it is not possible in the present context to neglect the LR terms, with $j \neq i$. Based on the considerations above, we hence, for observation points $(\vec{r}_i)$ inside molecule number $i$, replace Eq.(43) by the simpler one

$$\vec{E}(\vec{r}_i) = \vec{E}^{\,ext}(\vec{r}_i) - i\mu_0\omega \sum_{j(\neq i)} \int \overset{\leftrightarrow T}{D_0}(\vec{r}_i - \vec{r}_j) \cdot \overset{\leftrightarrow j}{\sigma}(\vec{r}_j, \vec{r}_j\,') \cdot \vec{E}(\vec{r}_j\,')d^3r_j'd^3r_j$$

$$-i\mu_0\omega \int \left[\overset{\leftrightarrow T}{g}(\vec{r}_i - \vec{r}_i\,') + \overset{\leftrightarrow L}{g}(\vec{r}_i - \vec{r}_i\,')\right] \cdot \overset{\leftrightarrow i}{\sigma}(\vec{r}_i\,', \vec{r}_i\,'') \cdot \vec{E}(\vec{r}_i\,'')d^3r_i''d^3r_i' \ ,$$

$$(44)$$

where coordinates inside molecule number $j$ have been given a subscript $j$.

Since the internal electrodynamics of the molecule is dominated by the self-coupling mechanism, Eq.(44) is solved as follows. The sum of the external field and the field stemming from all other molecules than the one in consideration, viz.

$$\vec{E}^B(\vec{r}_i) = \vec{E}^{\,ext}(\vec{r}_i)$$

$$- i\mu_0\omega \sum_{j(\neq i)} \int \overset{\leftrightarrow T}{D_0}(\vec{r}_i - \vec{r}_j) \cdot \overset{\leftrightarrow j}{\sigma}(\vec{r}_j, \vec{r}_j\,') \cdot \vec{E}(\vec{r}_j\,')d^3r_j'd^3r_j \ , \quad (45)$$

we name the background (B) field, and despite the fact that it depends on the unknown local field inside the various $j$-molecules, except from molecule number $i$, we assume in a first approximation that $\vec{E}^B(\vec{r}_i)$ is a *prescribed* field. Then, the unknown $\beta_{m_i n_i}$'s of the resulting equation

$$\vec{E}(\vec{r}_i) = \vec{E}^B(\vec{r}_i) + \sum_{m_i, n_i} \beta_{m_i n_i} \vec{F}^{SR}_{n_i m_i}(\vec{r}_i) \tag{46}$$

can be obtained in the standard manner from the matrix problem

$$\beta_{m_i n_i} - \sum_{o_i, p_i} N^{m_i n_i}_{o_i p_i}(SR)\beta_{o_i p_i} = H^B_{m_i n_i} \,, \tag{47}$$

where

$$H^B_{m_i n_i} = \int \vec{j}_{m_i n_i}(\vec{r}_i) \cdot \vec{E}^B(\vec{r}_i) d^3 r_i \,. \tag{48}$$

In the context of the subsequent (section 3) conceptual analysis of the field confinement problem for a mesoscopic particle (molecule) I shall discuss the solution of the self-coupling scheme further. For the time being it is sufficient to use the fact that the solution may be written in the form

$$\vec{E}(\vec{r}_i) = \int \overleftrightarrow{\Gamma}^i(\vec{r}_i, \vec{r}_i') \cdot \vec{E}^B(\vec{r}_i') d^3 r_i' \,, \tag{49}$$

where $\overleftrightarrow{\Gamma}^i(\vec{r}_i, \vec{r}_i')$ is the nonlocal field-field response tensor of molecule number $i$. By inserting Eq.(49), with $i = j$, into Eq.(45) one obtains

$$\vec{E}^B(\vec{r}_i) = \vec{E}^{ext}(\vec{r}_i)$$

$$-i\mu_0\omega \sum_{j(\neq i)} \int \overleftrightarrow{D}^T_0(\vec{r}_i - \vec{r}_j) \cdot \overleftrightarrow{\sigma}^j_B(\vec{r}_j, \vec{r}_j') \cdot \vec{E}^B(\vec{r}_j') d^3 r_j' d^3 r_j \,, \tag{50}$$

where

$$\int \overleftrightarrow{\sigma}^j_B(\vec{r}_j, \vec{r}_j') = \int \overleftrightarrow{\sigma}^j(\vec{r}_j, \vec{r}_j'') \cdot \overleftrightarrow{\Gamma}^j(\vec{r}_j'', \vec{r}_j') d^3 r_j'' \tag{51}$$

is the so-called background (B) conductivity tensor [7]. Physically, the background conductivity tensor relates in a nonlocal and linear fashion the current density $(\vec{J}^j(\vec{r}_j))$ inside a given molecule $(j)$ to the background field $(\vec{E}^B(\vec{r}_j))$ prevailing in the same molecule, i.e.

$$\vec{J}^j(\vec{r}_j) = \int \overleftrightarrow{\sigma}^j_B(\vec{r}_j, \vec{r}_j') \cdot \vec{E}^B(\vec{r}_j') d^3 r_j' \,. \tag{52}$$

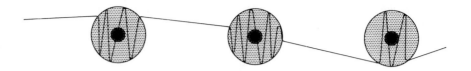

*Figure 3.* Schematic figure showing (i) the rapid variation of the local field in the region (grey domain) around a molecule (black domain) where the short range interactions are present, and (ii) the spatially slow variation of the transverse background field from molecule to molecule.

For optically dilute molecular media Eq.(50) forms a good starting point for practical local-field calculations, and constitutes a rigorous foundation for the point-dipole model's often used in molecular optics.

### 2.2.4. *The old and a new point-particle model in near field optics*

Since the transverse vacuum propagator $\overset{\leftrightarrow T}{D_0}(\vec{r}_i - \vec{r}_j)$ is slowly varying in space on the length scale of most molecules, it appears from Eq.(50) that also the background field is slowly varying across the individual molecules, see also Fig.3. If one completely neglects the variations in $\overset{\leftrightarrow T}{D_0}$ and $\vec{E}^B$ across the molecular domain, Eq.(50) is reduced to the form

$$\vec{E}^B(\vec{R}_i) = \vec{E}^{ext}(\vec{R}_i) + \sum_{j(\neq i)} \overset{\leftrightarrow}{T}(\vec{R}_i, \vec{R}_j) \cdot \vec{E}^B(\vec{R}_j) , \qquad (53)$$

with

$$\overset{\leftrightarrow}{T}(\vec{R}_i, \vec{R}_j) = -i\mu_0\omega \, \overset{\leftrightarrow T}{D_0}(\vec{R}_i - \vec{R}_j) \cdot \int \overset{\leftrightarrow j}{\sigma}_B(\vec{r}_j, \vec{r}'_j) d^3 r'_j d^3 r_j . \qquad (54)$$

To identify the positions of the various molecules, we have introduced the position (center of mass) coordinates $\vec{R}_i$ and $\vec{R}_j$. In terms of the electric dipole-electric dipole (ED-ED) polarizability [7]

$$\overset{\leftrightarrow j}{\alpha}(\omega) = \frac{i}{\omega} \int \overset{\leftrightarrow j}{\sigma}_B(\vec{r}_j, \vec{r}'_j) d^3 r'_j d^3 r_j , \qquad (55)$$

Eq.(54) can be written

$$\overset{\leftrightarrow}{T}(\vec{R}_i, \vec{R}_j) = -\mu_0\omega^2 \overset{\leftrightarrow T}{D_0}(\vec{R}_i - \vec{R}_j)\cdot \overset{\leftrightarrow j}{\alpha}(\omega) \ . \tag{56}$$

Together, Eqs.(53) and (56) constitute a rigorous basis for electric point-dipole descriptions of the electrodynamics in optically dilute molecular media.

In recent years, an electric point-dipole model closely related to the one introduced above has often been employed in studies in near-field optics[8-12]. This hitherto used model deviates from the present one in the sense that the coupling tensor $\overset{\leftrightarrow}{T}(\vec{R}_i, \vec{R}_j)$ is different. In the old (OLD) electric point-dipole version a coupling tensor of the form

$$\overset{\leftrightarrow OLD}{T}(\vec{R}_i, \vec{R}_j) = -\mu_0\omega^2 \overset{\leftrightarrow}{D_0}(\vec{R}_i - \vec{R}_j)\cdot \overset{\leftrightarrow j}{\alpha_0}(\omega) \tag{57}$$

was used. Instead of the transverse vacuum propagator in Eq.(33) the standard vacuum propagator [2]

$$\overset{\leftrightarrow}{D_0}(\vec{r} - \vec{r}')$$

$$= \overset{\leftrightarrow T}{D_0}(\vec{r} - \vec{r}') + \overset{\leftrightarrow L}{g_0}(\vec{r} - \vec{r}')$$

$$= \frac{q_0}{4\pi i}\left\{ \frac{1}{iq_0 R}\left(\overset{\leftrightarrow}{U} - \vec{e}_{\vec{R}}\vec{e}_{\vec{R}}\right) - \left[\frac{1}{(iq_0 R)^2} - \frac{1}{(iq_0 R)^3}\right]\left(\overset{\leftrightarrow}{U} - 3\vec{e}_{\vec{R}}\vec{e}_{\vec{R}}\right)\right\} e^{iq_0 R} \ , \tag{58}$$

or just its electrostatic part appears, and instead of the local-field dressed ED-ED polarizability, $\overset{\leftrightarrow j}{\alpha}(\omega)$, the bare polarizability

$$\overset{\leftrightarrow j}{\alpha_0}(\omega) = \frac{i}{\omega}\int \overset{\leftrightarrow j}{\sigma}(\vec{r}_j, \vec{r}'_j)d^3 r'_j d^3 r_j \ , \tag{59}$$

enters. By inserting Eq.(23) into Eq.(59) one obtains

$$\overset{\leftrightarrow j}{\alpha_0}(\omega) =$$

$$\frac{2}{(\hbar\omega)^2}\sum_{m_j, n_j}\left(\frac{\hbar\omega}{\varepsilon_{n_j} - \varepsilon_{m_j}}\right)\frac{(f_{n_j} - f_{m_j})(\varepsilon_{n_j} - \varepsilon_{m_j})^2}{\hbar\omega + \varepsilon_{m_j} - \varepsilon_{n_j}}\vec{P}_{n_j m_j}\vec{P}_{m_j n_j} \ , \tag{60}$$

where

$$\vec{P}_{\alpha\beta} = -e \int \psi_\beta^*(\vec{r})\vec{r}\psi_\alpha(\vec{r})d^3r \tag{61}$$

is the electric-dipole transition matrix element for excitation from state $\alpha$ to state $\beta$, $-e$ being the electron charge. Omitting the "diamagnetic" factor $\hbar\omega/(\varepsilon_{n_j} - \varepsilon_{m_j})$, the expression for $\overset{\leftrightarrow}{\alpha}_0^j$ can be recognized as the standard expression for the bare (local-limit) paramagnetic polarizability [13]. The bare and dressed ED-ED polarizabilities are related via

$$\overset{\leftrightarrow}{\alpha}^j(\omega) = \overset{\leftrightarrow}{\alpha}_0^j(\omega) + \overset{\leftrightarrow}{\Delta}^j(\omega) , \tag{62}$$

where $\overset{\leftrightarrow}{\Delta}^j(\omega)$ is the local-field correction. This correction is associated with the short range interactions mentioned previously, and has been studied in e.g. Refs.[7,14,15].

The old ( $\overset{\leftrightarrow}{T}^{OLD}$ ) and new ( $\overset{\leftrightarrow}{T}$ ) coupling tensors deviate conceptually from each other. The new one has been obtained by a rigorous derivation starting from the microscopic local-field theory, whereas the old one was introduced in a heuristic manner. In the old point-dipole theory short range effects are to some extent included in the propagator ($\overset{\leftrightarrow}{D}_0$), but since the transverse part of the short range dynamics, which in the rigorous theory appears via the self-field propagator $\overset{\leftrightarrow}{g}^T$ , is omitted, it is doubtful whether one can rely on the use of $\overset{\leftrightarrow}{T}^{OLD}$ in the near field ($\sim R^{-3}$) zone. In the context of near field optical calculations where only the quasi-static near field part of $\overset{\leftrightarrow}{T}^{OLD}(\vec{R})$, i.e. precisely $\overset{\leftrightarrow}{g}^L(\vec{R})$, is used to study e.g. the interaction between a probe tip and surface particles (or between the particles themselves), one should be careful not to rely too much on the obtained results (conclusions). In the rigorous (new) point-dipole approach longitudinal as well as transverse short range effects are included not in the electromagnetic propagator but in the dressed polarizability, $\overset{\leftrightarrow}{\alpha}^j(\omega)$. The inclusion of SR effects in $\overset{\leftrightarrow}{\alpha}^j(\omega)$ of course means that these must be excluded from the propagator, hence the presence of $\overset{\leftrightarrow}{D}_0^T(\vec{R})$ instead of $\overset{\leftrightarrow}{D}_0(\vec{R})$ in our new theory. As justified by a number of local-field studies in mesoscopic media [2], and further discussed in section 3, it is important to keep the interplay between the transverse and longitudinal electrodynamics in the short range interaction zone. A further bonus of the new point-dipole theory is that it has become apparent that only when the molecular system is so dilute that optical short range interactions between the individual (neighbouring) molecules are negligible it is completely safe to introduce a point-dipole model.

Before going on, I would like to emphasize that though I have limited myself here to electric-dipole (ED) absorption/radiation theory, it is certainly possible to extend the rigorous point-particle theory so as to take into account electric-quadrupole (EQ), magnetic-dipole (MD), and if needed higher order multipole absorption and radiation effects. Basically, one just needs to Taylor expand $\overleftrightarrow{D}_0^T(\vec{r}_j - \vec{r}_j)$ and $\vec{E}^B(\vec{r}'_j)$ appearing under the integral sign(s) of Eq.(50) beyond zeroth order [2].

### 2.2.5. A new macroscopic theory and comments on the old macroscopic approach

For optically dilute molecular systems it is possible to include short range electrodynamic interactions in the polarizability of the individual molecules and thereby establish in a rigorous manner a point-particle model enabling one to study long range local-field effects in these systems. The long range part of the local field we have named the background field $(\vec{E}^B(\vec{r}))$, and its values taken at the various molecular positions satisfy the following algebraic equation:

$$\vec{E}^B(\vec{R}_i) = \vec{E}^{ext}(\vec{R}_i) - \mu_0\omega^2 \sum_{j(\neq i)} \overleftrightarrow{D}_0^T(\vec{R}_i - \vec{R}_j)\cdot \overleftrightarrow{\alpha}^j(\omega)\cdot \vec{E}^B(\vec{R}_j) . \quad (63)$$

Letting $i$ run through all molecular numbers, Eq.(63) generates a set of algebraic equations from which the background (long range) field prevailing on each of the interacting molecules may be obtained. If the system in consideration contains a huge number of molecules in a volume of linear extension comparable to the length scale over which the transverse vacuum propagator varies, Eq.(63) may be replaced by an integral relation, viz.

$$\vec{E}^B(\vec{r}) = \vec{E}^{ext}(\vec{r}) - \mu_0\omega^2 \int \overleftrightarrow{D}_0^T(\vec{r} - \vec{r}')\cdot \overleftrightarrow{\alpha}(\vec{r}';\omega)\cdot \vec{E}^B(\vec{r}')d^3r' , \quad (64)$$

where $\overleftrightarrow{\alpha}(\vec{r}';\omega)$ plays the role of a position-dependent dressed polarizability. The integral equation in (64) to my knowledge is a new one, and it constitutes a rigorous basis for macroscopic electrodynamic studies in optically dilute media. In the present context the external field is always divergence-free, $\vec{E}^{ext} = \vec{E}_T^{ext}$. This, together with the fact that only the transverse propagator, $\overleftrightarrow{D}_0^T$, appears in Eq.(63) (and Eq.(64)) immediately implies that the background field has no rotational-free component, i.e. $\vec{E}^B = \vec{E}_T^B$. The algebraic equation in (63) thus takes the form

$$\vec{E}_T^B(\vec{R}_i) = \vec{E}_T^{ext}(\vec{R}_i) - \mu_0\omega^2 \sum_{j(\neq i)} \overset{\leftrightarrow T}{D_0}(\vec{R}_i - \vec{R}_j) \cdot \overset{\leftrightarrow j}{\alpha}(\omega) \cdot \vec{E}_T^B(\vec{R}_j) . \quad (65)$$

On the basis of the position-dependent dressed polarizability $\overset{\leftrightarrow}{\alpha}(\vec{r}';\omega)$, which essentially is a macroscopic quantity, we can introduce a macroscopic relative dielectric tensor, $\overset{\leftrightarrow}{\varepsilon}(\vec{r};\omega)$, via the standard relation

$$\overset{\leftrightarrow}{\varepsilon}(\vec{r};\omega) = \overset{\leftrightarrow}{U} + \varepsilon_0^{-1} \overset{\leftrightarrow}{\alpha}(\vec{r};\omega) . \quad (66)$$

By means of Eq.(66) one may rewrite the macroscopic integral equation for the transverse background field as follows:

$$\vec{E}_T^B(\vec{r};\omega) = \vec{E}_T^{ext}(\vec{r};\omega)$$
$$-(\frac{\omega}{c_0})^2 \int \overset{\leftrightarrow T}{D_0}(\vec{r} - \vec{r}';\omega) \cdot \left( \overset{\leftrightarrow}{\varepsilon}(\vec{r}';\omega) - \overset{\leftrightarrow}{U} \right) \cdot \vec{E}_T^B(\vec{r}';\omega) d^3 r' , \quad (67)$$

stressing also the frequency dependence of the various quantities. Let us now compare the rigorously established macroscopic integral equation in Eq.(67) with the old one used previously by many researchers in e.g. near field optics [1]. For conceptual clarity we add a superscript *macro* to the various quantities of the old integral equation, which hence reads

$$\vec{E}^{macro}(\vec{r}) = \vec{E}^{ext}(\vec{r})$$
$$-(\frac{\omega}{c_0})^2 \int \overset{\leftrightarrow}{D_0}(\vec{r} - \vec{r}') \cdot \left( \overset{\leftrightarrow macro}{\varepsilon}(\vec{r}') - \overset{\leftrightarrow}{U} \right) \cdot \vec{E}^{macro}(\vec{r}') d^3 r' ,$$

$$(68)$$

where the macroscopic dielectric tensor $\overset{\leftrightarrow macro}{\varepsilon}(\vec{r}';\omega)$ is related to the bare macroscopic polarizability $\overset{\leftrightarrow}{\alpha}_0(\vec{r}';\omega)$. In the old macroscopic formulation one thus deals with a mixture of long and short range interactions witch is conceptually unclear and not rigorous cf. the discussion of $\overset{\leftrightarrow}{T}(\vec{R}_i, \vec{R}_j)$ and $\overset{\leftrightarrow OLD}{T}(\vec{R}_i, \vec{R}_j)$ in section 2.2.4. In using $\overset{\leftrightarrow}{D_0}(\vec{r} - \vec{r}')$ in Eq.(68), the transverse part, $\overset{\leftrightarrow T}{g}(\vec{r} - \vec{r}')$, of the self-field propagator is omitted. In some of the macroscopic studies transverse self-field effects are taken into account in an approximate manner, i.e. by replacing the transverse delta function

$\overset{\leftrightarrow T}{\delta}(\vec{r} - \vec{r}\,')$ appearing in Eq.(31) by the Dirac delta function itself (times the unit tensor), i.e.[16]

$$\overset{\leftrightarrow T}{g}(\vec{r} - \vec{r}\,') \Longrightarrow \frac{1}{3}(\frac{c_0}{\omega})^2 \delta(\vec{r} - \vec{r}\,')\overset{\leftrightarrow}{U} \ .$$

Using such an extension, the macroscopic integral equation in (68) is replaced by the following one [16]

$$\vec{E}^{macro}(\vec{r}) = \vec{E}^{ext}(\vec{r}) - \frac{1}{3}\left(\overset{\leftrightarrow macro}{\varepsilon}(\vec{r}) - \overset{\leftrightarrow}{U}\right) \cdot \vec{E}^{macro}(\vec{r})$$

$$-(\frac{\omega}{c_0})^2 \int \overset{\leftrightarrow}{D}_0(\vec{r} - \vec{r}\,') \cdot \left(\overset{\leftrightarrow macro}{\varepsilon}(\vec{r}\,') - \overset{\leftrightarrow}{U}\right) \cdot \vec{E}^{macro}(\vec{r}\,')d^3r' \ ,$$

$$\tag{69}$$

or the equivalent one

$$\frac{1}{3}\left(\overset{\leftrightarrow macro}{\varepsilon}(\vec{r}) + 2\overset{\leftrightarrow}{U}\right) \cdot \vec{E}^{macro}(\vec{r}) =$$

$$\vec{E}^{ext}(\vec{r}) - (\frac{\omega}{c_0})^2 \int \overset{\leftrightarrow}{D}_0(\vec{r} - \vec{r}\,') \cdot \left(\overset{\leftrightarrow macro}{\varepsilon}(\vec{r}\,') - \overset{\leftrightarrow}{U}\right) \cdot \vec{E}^{macro}(\vec{r}\,')d^3r' \ .$$

$$\tag{70}$$

Even though transverse short range interactions now at least approximately have been included, the rigorously derived Eq.(67) and the heuristically established Eq.(70) are still different, the main difference stemming from the fact that one tries to share the short range interactions between the propagator and the dielectric function in the standard theory. From a rigorous point of view it seems impossible to achieve such a goal, and the physical interpretation becomes unclear.

Let us briefly consider the case where the optically dilute molecular medium from a macroscopic point of view is homogeneous and isotropic. Thus, with the replacement $\overset{\leftrightarrow macro}{\varepsilon}(\vec{r}; \omega) \Longrightarrow \varepsilon^{macro}(\omega)\overset{\leftrightarrow}{U}$, Eq.(70) becomes

$$\frac{\varepsilon^{macro} + 2}{3}\vec{E}^{macro}(\vec{r}) = \vec{E}^{ext}(\vec{r})$$

$$-(\frac{\omega}{c_0})^2(\varepsilon^{macro} - 1) \int \overset{\leftrightarrow}{D}_0(\vec{r} - \vec{r}\,') \cdot \vec{E}^{macro}(\vec{r}\,')d^3r' \ . \tag{71}$$

If in this equation one neglects by *brute force* the longitudinal electrodynamics, as one often does in optics, and thus use the restriction $\vec{\nabla} \cdot \vec{E}^{macro} = \vec{\nabla} \cdot \vec{E}_T^{macro} = 0$, Eq.(71) is reduced to

$$\frac{\varepsilon^{macro} + 2}{3} \vec{E}_T^{\,macro}(\vec{r}) = \vec{E}_T^{\,ext}(\vec{r})$$

$$-(\frac{\omega}{c_0})^2(\varepsilon^{macro} - 1) \int \overleftrightarrow{D}_0^T(\vec{r} - \vec{r}') \cdot \vec{E}_T^{\,macro}(\vec{r}')d^3r' \, , \qquad (72)$$

an equation which in fact is a bit closer to the rigorous macroscopic equation (67) [with $\overleftrightarrow{\varepsilon}(\vec{r}';\omega) = \varepsilon(\omega) \overleftrightarrow{U}$ ]. In the nonretarded ($c_0 \to \infty$) limit the macroscopic and external fields are related via the textbook result $\vec{E}^{\,macro}(\vec{r}) = 3[\overleftrightarrow{\varepsilon}^{macro}(\vec{r}) + 2 \overleftrightarrow{U}]^{-1} \cdot \vec{E}^{\,ext}(\vec{r})$ in the nonrigorous approach, whereas $\vec{E}_T^B = \vec{E}_T^{ext}$ in the rigorous theory.

In finishing this section let me make a comment on the computational efforts needed when using respectively the new (Eq.(67)) and old (Eq.(68)) macroscopic integral equation. In the new one the main numerical challenge is associated with the calculation of the dressed dielectric tensor $\overleftrightarrow{\varepsilon}(\vec{r};\omega)$ (from the dressed polarizability $\overleftrightarrow{\alpha}^j(\omega)$ of the individual molecules), whereas the main challenge for the old type of calculation arises from the fact that the $R^{-3}$-singularity in $\overleftrightarrow{D}_0$ makes the volume integral only conditionally convergent. Since the singularity in $\overleftrightarrow{D}_0^T(\vec{R})$ essentially is of the $R^{-1}$-type the volume integral in Eq.(67) is absolutely convergent.

## 3. On the spatial field confinement problem and its relation to near field optics

As already pointed out in the introduction to section 2, it is major goal in near field optics to increase the spatial resolution towards the atomic limit. Whether or not this will be possible is an open question at the time of writing, but it is obvious that a new generation of experiments, as well as a reexamination of the theoretical foundation, are needed. In the present section first I shall with a critical eye briefly review the framework and shortcomings of existing theoretical approaches aiming at examining the limit for the spatial resolution. Then, I shall present what appears to be a new view on the resolution problem. In order for the reader to grasp the main points of the new formalism, I shall limit myself to a conceptual discussion of the spatial confinement problem for the electromagnetic field.

### 3.1. HEURISTIC APPROACHES AND THEIR SHORTCOMINGS

In textbook expositions it is usually claimed that the spatial resolution in optics is limited by the Rayleigh criterion. For a slit of width $a$ and monochromatic light of wavelength $\lambda_0$, the minimum angular separation $\theta_{min}$ subtended by the source at the slit thus is given by $\theta_m \simeq \sin\theta_m = \lambda_0/a$. The Rayleigh criteria is based on the Fraunhofer diffraction pattern, which one calculates using the *far field* scalar Green's function $exp(iq_0 \mid \vec{r} - \vec{r}' \mid) / (4\pi \mid \vec{r} - \vec{r}' \mid)$. Not only in Fraunhofer diffraction but also in Fresnel diffraction one takes as a starting point the far field propagator. This means that in so-called classical optics statements about the spatial resolution necessarily break down on a length scale so small that the far field propagator cannot describe the field propagation accurately. According to e.g. Eq.(33) this happens when $\mid \vec{r} - \vec{r}' \mid /\lambda_0 \stackrel{<}{\sim} (2\pi)^{-1}$, i.e. roughly speaking when the involved distances become one to ten times smaller than the optical (vacuum) wavelength.

In near field optics one aims at carrying out optical microscopy beyond the classical diffraction limit. From a theoretical point of view, this necessitates that middle and near field propagation effects of light are considered. In the standard theory these effects are treated by means of the (dyadic) vacuum propagator $\overleftrightarrow{D}_0 = \overleftrightarrow{D}_0^T + \overleftrightarrow{g}^L$, which explicit expression is given in Eq.(58). Despite the fact that this propagator exhibits a strong $R^{-3}$-singularity, a number of useful information may be obtained starting from this propagator. Thus, it has for instance been shown [12] that so-called configurational resonances often may be studied using the $\overleftrightarrow{D}_0$-propagator. The configurational resonances are of interest in the present context because these among other things show that the strongest coupling between the probe tip and the particles of the surface under study occurs at a finite dipole-dipole distance, though the dipole-dipole interaction diverges as $R^{-6}$ when the point dipoles approach each other.

Although the point-particle model certainly is a valuable theoretical tool for optical near field microscopy, it is not obvious that this model can be used to discuss the possibilities of reaching a spatial resolution on the atomic length scale. After all, atoms (molecules) are not point-particles! In the self-energy quantum electrodynamics of atoms, and in local-field electrodynamics of quantum dots (and small particles) it is known that the point-particle model must be abandoned since it gives rise to unphysical divergences. Taking into account the finite size of the particle region in question (single atom, molecule, mesoscopic particle, fibre tip of a near field optical microscope) one might be tempted to believe that the confinement region for the field essentially coincides with the spatial domain occupied by the electrons of the particle considered. In quantum mechanical terms, the particle region is given by the (many-body) electron density in the field-unperturbed state. On second thoughts it is realized that because all interactions in electrodynamics (and hence in near field optics) involve a finite probability for transferring the system from its ground state to one

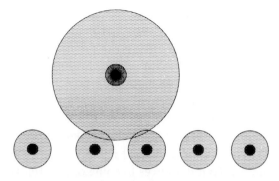

*Figure 4.* Schematic figure illustrating the near field interaction of a probe tip molecule (big black domain) with an array of surface molecules. When the short range forces, the extension of which are given by the grey domains, tend to overlap the optical near field resolution is gradually lost.

or more of its excited states, it appears more correct to identify the spatial domain in question with the domain occupied by the ground state and all the states to which excitation takes place in the actual case. Rigorously speaking, the confinement region for the electrons in optics thus coincides with the spatial domain occupied by all the relevant transition current densities. In the one-electron approximation the individual transition current densities are given by Eq.(9). It seems conceptually clear however that the question of the spatial resolution in near field optics is not directly related to the region occupied by the transition current densities of relevance, that is, one may very well loose the spatial resolution before the transition current densities of the (two) particles one wants to resolve start to overlap. What matters in near field optics is the spatial confinement of the electromagnetic field accompanying the relevant transition current densities of our particle. Hence, we expect that two particles in near field optics spatially cannot be resolved if the attached fields accompanying the transition current densities of two particles overlap appreciably (see Fig.4).

## 3.2. CONFINEMENT OF ATTACHED FIELDS

Let us consider the situation where a molecular probe tip interacts electromagnetically with a number of molecules, the spatial positions of which we want to determine (resolve). Instead of molecules we might of course think of atoms or mesoscopic particles. We assume from the outset that the system of molecules is so dilute that there is no short range interactions among the different molecules (including the tip molecule). As we shall realize *post priori*, this assumption relates directly to the very spatial resolution problem. From an orthodox point of view, one would in this case naturally start the analysis from the phenomenological point-particle model. However, as we have realized in section 2.2, a rigorous starting point is Eq.(44). In order to study the spatial confinement of the attached fields, we leave out the far field coupling between the individual molecules from the dynamics. Hence, to determine the local field inside and in the vicinity of a particular molecule, say number $i$, we start from the integral equation

$$\vec{E}(\vec{r}_i) = \vec{E}^B(\vec{r}_i)$$
$$-i\mu_0\omega \int [\overset{\leftrightarrow}{g}^T(\vec{r}_i - \vec{r}'_i) + \overset{\leftrightarrow}{g}^L(\vec{r}_i - \vec{r}'_i)] \cdot \overset{\leftrightarrow}{\sigma}^i(\vec{r}'_i, \vec{r}''_i) \cdot \vec{E}(\vec{r}''_i) d^3 r''_i d^3 r'_i .$$

$$(73)$$

By merely looking at the expression for the self-field propagators, $\overset{\leftrightarrow}{g}^L(\vec{R}_i)$ and $\overset{\leftrightarrow}{g}^T(\vec{R}_i)$, in Eq.(34), one might be tempted to believe that the spatial range of the attached field, $\vec{E}(\vec{r}_i)$, is comparable to the "spatial extension" of these propagators. From Eq.(34) it appears that this extension is precisely the range of the near field zone of the standard propagator $\overset{\leftrightarrow}{D}_0$, see Eq.(58). If this was the answer to the spatial resolution problem, one would not be able to improve beyond the $(q_0 R)^{-3}$-tail extension i.e. beyond the point where the near field part of $\overset{\leftrightarrow}{D}_0$ is the dominating part. In turn this means that "near field optics" would really be the art of middle field optics, and nothing more! In passing I would like to emphasize here that although many investigations are classified as belonging to near field optics, the observations are mainly carried out in the middle field zone.

Before proceeding along the main line of reasoning, let us briefly consider the type of confinement one ends up with starting with the point-particle model. For a point-particle (number $i$) located at the site $\vec{R}_i$, the induced current density is given by

$$\vec{J}^i(\vec{r}'_i) = \delta(\vec{r}'_i - \vec{R}_i) \overset{\leftrightarrow}{S}_0^i \cdot \vec{E}^B(\vec{R}_i) , \qquad (74)$$

where

$$\overset{\leftrightarrow}{S}_0^i = \int \overset{\leftrightarrow}{\sigma}^i(\vec{r}'_i, \vec{r}''_i) d^3 r''_i d^3 r'_i = -i\omega \overset{\leftrightarrow}{\alpha}_0^j \qquad (75)$$

is the conductivity tensor of our electric point-dipole. By combining Eqs.(73) and (74), one obtains

$$\vec{E}(\vec{r}_i) = \vec{E}^B(\vec{r}_i) - i\mu_0\omega[\overleftrightarrow{g}^T(\vec{r}_i - \vec{R}_i) + \overleftrightarrow{g}^L(\vec{r}_i - \vec{R}_i)] \cdot \overleftrightarrow{S}_0^i \cdot \vec{E}^B(\vec{R}_i) , \quad (76)$$

or with the use of Eq.(34)

$$\vec{E}(\vec{r}_i) = \vec{E}^B(\vec{r}_i)$$
$$-\frac{\mu_0\omega q_0}{6\pi} \frac{1}{(iq_0 | \vec{r}_i - \vec{R}_i |)^3} \left( \overleftrightarrow{U} - 3\vec{e}_{\vec{r}_i - \vec{R}_i} \vec{e}_{\vec{r}_i - \vec{R}_i} \right) \cdot \overleftrightarrow{S}_0^i \cdot \vec{E}^B(\vec{R}_i) . \quad (77)$$

Within the framework of the point-dipole model one thus, as expected, ends up with a $\vec{R}^{-3}$-confinement exhibiting a strong singularity at the position of the point particle.

Returning to the microscopic approach described in section 2 (and in particular in section 2.2.3) we know that the attached field inside and in the vicinity of the molecule in question is given by Eq.(46). Inserting in this equation the expression in Eq.(38) for the short range part of $\vec{F}_{n_i m_i}$ one obtains, leaving out the index reference to the molecule number from the notation

$$\vec{E}(\vec{r}) = \vec{E}^B(\vec{r}) + \sum_{m,n} W_{mn} \left( \vec{j}^L_{nm}(\vec{r}) + \frac{1}{3}\vec{j}^T_{nm}(\vec{r}) \right) , \quad (78)$$

where

$$W_{mn} = \frac{2\beta_{mn}}{\varepsilon_0\omega^2} \frac{\hbar\omega}{\varepsilon_n - \varepsilon_m} \frac{f_m - f_n}{\hbar\omega + \varepsilon_m - \varepsilon_n} . \quad (79)$$

It appears from Eq.(78) that the confinement region for the field belonging to a given transition, e.g. the $m \Leftrightarrow n$ transition, is given by the extension of the longitudinal (or transverse) part of the transition current density belonging to this transition. The weight factor which is $W_{mn}$ for $\vec{j}^L_{nm}$ ( and $\frac{1}{3}W_{mn}$ for $\vec{j}^T_{nm}$ ) depends via $\beta_{mn}$ on the coupling between the various transitions.

To gain further insight in the confinement problem it is useful to take a closer look at the relations between the transverse and longitudinal parts of the transition current density and the total transition current density. Thus, starting from the relations

$$\vec{j}^T_{nm}(\vec{r}) = \vec{\nabla} \times \left[ \vec{\nabla} \times \int \frac{\vec{j}_{nm}(\vec{r}')d^3r'}{4\pi | \vec{r} - \vec{r}' |} \right] , \quad (80)$$

$$\vec{j}^{\,L}_{nm}(\vec{r}) = -\vec{\nabla}\left[\vec{\nabla}\cdot\int\frac{\vec{j}_{nm}(\vec{r}\,')d^3r'}{4\pi\,|\,\vec{r}-\vec{r}\,'\,|}\right]\,, \tag{81}$$

it is possible to rewrite the transverse and longitudinal current densities in a more convenient manner, that is [7,16]

$$\vec{j}^{\,T}_{nm}(\vec{r}) = \frac{2}{3}\vec{j}_{nm}(\vec{r}) + \vec{\mathcal{J}}_{nm}(\vec{r}) \tag{82}$$

$$\vec{j}^{\,L}_{nm}(\vec{r}) = \frac{1}{3}\vec{j}_{nm}(\vec{r}) - \vec{\mathcal{J}}_{nm}(\vec{r})\,, \tag{83}$$

where

$$\vec{\mathcal{J}}_{nm}(\vec{r}) = PV\int\frac{(3\vec{e}_{\vec{r}-\vec{r}\,'}\vec{e}_{\vec{r}-\vec{r}\,'} - \overleftrightarrow{U})\cdot\vec{j}_{nm}(\vec{r}\,')}{4\pi\,|\,\vec{r}-\vec{r}\,'\,|^3}d^3r'\,. \tag{84}$$

In Eq.(84), PV stands for principal value, which in the present context means that the conditionally convergent integral has to be performed with a small sphere of radius $\varepsilon$ centered on $\vec{r}$ excluded from the domain of integration. After having carried out the integration one takes the limit $\varepsilon \to 0$. By inserting Eqs.(82) and (83) into Eq.(78) the local field becomes

$$\vec{E}(\vec{r}) = \vec{E}^B(\vec{r}) + \sum_{m,n} W_{mn}\left(\frac{5}{9}\vec{j}_{nm}(\vec{r}) - \frac{2}{3}\vec{\mathcal{J}}_{nm}(\vec{r})\right)\,. \tag{85}$$

Outside the spatial domain occupied by the electronic current densities themselves, i.e. the $\vec{j}_{nm}$'s, the attached part of the local field is given by the $\vec{\mathcal{J}}_{nm}$'s. The physical meaning of $\vec{\mathcal{J}}_{nm}$ readily appears if one divides $\vec{\mathcal{J}}_{nm}$ by the factor $i\varepsilon_0\omega$. Thus, the quantity

$$\frac{\vec{\mathcal{J}}_{nm}(\vec{r})}{i\varepsilon_0\omega} = \frac{i}{\omega}PV\int\frac{\left(\overleftrightarrow{U} - 3\vec{e}_{\vec{r}-\vec{r}\,'}\vec{e}_{\vec{r}-\vec{r}\,'}\right)\cdot\vec{j}_{nm}(\vec{r}\,')d^3r'}{4\pi\varepsilon_0\,|\,\vec{r}-\vec{r}\,'\,|^3} \tag{86}$$

may readily be recognized as the quasi-static expression for the electric field at $\vec{r}$ stemming from a continuous distribution of electric dipoles having the density $i\vec{j}_{nm}(\vec{r}\,')/\omega$. For observation points far from the electric-dipole distribution, i.e. for $|\,\vec{r}\,| \to \infty$, and provided the quantity $\int\vec{j}_{nm}(\vec{r}\,')d^3r'$ is different from zero the field is given by the asymptotic expression

$$\frac{\vec{\mathcal{J}}^{\,\infty}_{nm}(\vec{r})}{i\varepsilon_0\omega} = \frac{i}{\omega}\frac{1}{4\pi\varepsilon_0 r^3}\left(\overleftrightarrow{U} - 3\vec{e}_{\vec{r}}\vec{e}_{\vec{r}}\right)\cdot\int\vec{j}_{nm}(\vec{r}\,')d^3r'\,, \tag{87}$$

assuming that the origo of the local coordinate system is placed at the centre of mass of the molecule in question. Since,

$$\frac{i}{\omega} \int \vec{j}_{nm}(\vec{r}\,')d^3r' = \frac{\varepsilon_n - \varepsilon_m}{\hbar\omega}\vec{P}_{nm} \ , \tag{88}$$

it is seen that

$$\frac{\vec{J}_{nm}^{\infty}(\vec{r})}{i\varepsilon_0\omega} = \frac{\varepsilon_n - \varepsilon_m}{\hbar\omega}\frac{1}{4\pi\varepsilon_0 r^3}\left(\overleftrightarrow{U} - 3\vec{e}_{\vec{r}}\vec{e}_{\vec{r}}\right) \cdot \vec{P}_{nm} \ , \tag{89}$$

apart from a factor $(\varepsilon_n - \varepsilon_m)/(\hbar\omega)$, equals the electrostatic expression for the electric field from an electric point-dipole of moment $\vec{P}_{nm}$. If the $n \Rightarrow m$ transition is electric-dipole forbidden, the asymptotic field is of the electric-quadrupole plus magnetic-dipole type.

By means of Eqs.(79) and (89), one obtains the following *implicit* expression for the asymptotic part of the attached field

$$-\frac{2}{3}\sum_{m,n}W_{mn}\vec{J}_{nm}^{\infty}(\vec{r}) = \frac{4}{3}\frac{1}{4\pi\varepsilon_0 r^3}\left(\overleftrightarrow{U} - 3\vec{e}_{\vec{r}}\vec{e}_{\vec{r}}\right)$$
$$\cdot\left\{\sum_{m,n}\frac{f_n - f_m}{\hbar\omega + \varepsilon_m - \varepsilon_n}\vec{P}_{nm}\left[\frac{i}{\omega}\int\vec{j}_{mn}(\vec{r}\,') \cdot \vec{E}(\vec{r}\,')d^3r'\right]\right\} \ . \tag{90}$$

If one assumes that the background field is constant across the molecular domain, Eq.(90) can be written in terms of the dressed polarizability, $\overleftrightarrow{\alpha}(\omega)(\equiv\overleftrightarrow{\alpha}^i(\omega))$, as follows:

$$-\frac{2}{3}\sum_{m,n}W_{mn}\vec{J}_{nm}^{\infty}(\vec{r}) = -\frac{2}{3}\frac{1}{4\pi\varepsilon_0 r^3}\left(\overleftrightarrow{U} - 3\vec{e}_{\vec{r}}\,\vec{e}_{\vec{r}}\right) \cdot \overleftrightarrow{\alpha}(\omega) \cdot \vec{E}^B(\vec{0}) \ , \tag{91}$$

letting the centre of mass of the molecule under consideration be located at the origo of our coordinate system. At this point it is interesting to compare the result in Eq.(91) with that obtained on the basis of the rigorous point-dipole approach. In terms of the bare polarizability $\overleftrightarrow{\alpha}_0(\omega)(\equiv\overleftrightarrow{\alpha}_0^i(\omega)) = (i/\omega)\overleftrightarrow{S}_0$, it appears from Eq.(77) that the attached electric dipole field is given by

$$\vec{E}(\vec{r}) - \vec{E}^B(\vec{0}) = -\frac{2}{3}\frac{1}{4\pi\varepsilon_0 r^3}\left(\overleftrightarrow{U} - 3\vec{e}_{\vec{r}}\,\vec{e}_{\vec{r}}\right) \cdot \overleftrightarrow{\alpha}_0(\omega) \cdot \vec{E}^B(\vec{0}) \ . \tag{92}$$

This result resembles that of the rigorous calculation, the only difference being that the dressing of the polarizability $\overset{\leftrightarrow}{\alpha}_0$ has to be omitted if one contracts the electronic current density distribution to a point before calculating the asymptotic part of the attached field. Note that if the old electric point-dipole model had been used the factor 2/3 would be missing.

If one retains only the asymptotic part of the attached field, given by Eq.(91), one may insist in using the point-dipole model with an $r^{-3}$-term. Hence, if a *single* molecule located at $\vec{r} = \vec{0}$ is subjected to an external field which is slowly varying across the molecule, the field in the asymptotic region and beyond is given by

$$
\vec{E}(\vec{r}) = \vec{E}^{ext}(\vec{r}) - \mu_0\omega^2 \left( \overset{\leftrightarrow T}{D_0}(\vec{r}) + \frac{2}{3} \overset{\leftrightarrow L}{g}(\vec{r}) \right) \cdot \overset{\leftrightarrow}{\alpha}(\omega) \cdot \vec{E}^{ext}(\vec{0})
$$

$$
= \vec{E}^{ext}(\vec{r}) - \mu_0\omega^2 \left( \overset{\leftrightarrow}{D_0}(\vec{r}) - \frac{1}{3} \overset{\leftrightarrow L}{g}(\vec{r}) \right) \cdot \overset{\leftrightarrow}{\alpha}(\omega) \cdot \vec{E}^{ext}(\vec{0}) , \quad (93)
$$

where $\overset{\leftrightarrow}{\alpha}(\omega)$ is the *dressed* polarizability.

## 3.3. CONTACT INTERACTIONS

Let me finish the analysis with a brief discussion of the short range matrix elements, $N_{op}^{mn}$, coupling the various pairs of electronic transitions inside a single molecule (molecule index omitted) together, electrodynamically. According to Eq.(41), the coupling is proportional to the integral

$$
\int \vec{j}_{mn}(\vec{r}) \cdot \left( \frac{1}{3} \vec{j}_{po}^{T}(\vec{r}) + \vec{j}_{po}^{L}(\vec{r}) \right) d^3r = \frac{5}{9} \int \vec{j}_{mn}(\vec{r}) \cdot \vec{j}_{po}(\vec{r}) d^3r
$$

$$
+ \frac{2}{3} PV \int \frac{1}{4\pi |\vec{r} - \vec{r}'|^3} \vec{j}_{mn}(\vec{r}) \cdot \left( \overset{\leftrightarrow}{U} - 3\vec{e}_{\vec{r}-\vec{r}'} \vec{e}_{\vec{r}-\vec{r}'} \right) \cdot \vec{j}_{po}(\vec{r}') d^3r' d^3r .
$$

$$
(94)
$$

The second term on the right hand side of Eq.(94), essentially, represents the dipole-dipole interaction between the two induced transition current density flows, $\vec{j}_{mn}(\vec{r})$ and $\vec{j}_{po}(\vec{r})$. The first term, in which the integrand only depends on the transition current densities at the *same* point $\vec{r}$, one may call the contact interaction term. Readers familiar with studies of the magnetic interaction between the nuclear spin and the electron spin in an atom will remember that a (similar) contact interaction between magnetic densities occurs in hyperfine-structure calculations.

## 4. Summary

Starting from microscopic electrodynamics a local-field theory for optically dilute molecular media has been established, and by separating short and long range electrodynamic interactions a new point-particle model, possibly useful in near field optics, is developed. It has been demonstrated that the phenomenologically based point-dipole model previously used has no rigorous basis in microscopic local-field theory. The spatial field confinement problem and its relation to near field optics are studied, and by dividing the so-called attached field into contact and dipolar parts, an asymptotic $R^{-3}$-term is identified. By means of a short range field dressed polarizability, and the adding of the $R^{-3}$-dipole tail to the transverse long range propagator a new point-dipole radiator model is introduced.

### Acknowledgement

Here, I wish to express my gratitude to Professor Manuel Nieto-Vesperinas for inviting me to speak at the Miraflores workshop and for creating the framework for a stimulating and fruitful meeting on near field optics.

### References

1. Pohl, D.W. and Courjon, D. eds. (1993) *Near Field Optics*, NATO ASI Series, Series E: Applied Sciences - Vol.242, Kluwer Academic Publishers, Dordrecht; and references herein.
2. Keller, O. (1995) Local Fields in the Electrodynamics of Mesoscopic Media, *Physics Reports* **535** (in press); and references herein.
3. Keller, O. (1995) Local-Field Studies in the Nonlinear Optics of Mesoscopic Systems, in O.Keller (ed.), *Studies in Classical and Quantum Nonlinear Optics*, Nova Science Publishers, New York, pp. 269-335.
4. Cohen-Tannoudji, C., Dupont-Roc, J. and Grynberg, G. (1989) *Photons and Atoms, Introduction to Quantum Electrodynamics*, Wiley-Interscience, New York.
5. Mahan, G.D. (1990) *Many-Particle Physics*, Plenum Press, New York.
6. Keller, O. (1996) Aspects of local-field electrodynamics in condensed matter, in A.Shumovsky (ed.), *Advances in Quantum Optics and Spectroscopy of Solids*, Springer-Verlag, Berlin (in press).
7. Keller, O. (1994) Optical polarizability of small quantum particles: local field effects in a self-field approach, *J. Opt. Soc. Amer.* **B11**, 1480-1489.
8. Labani, B., Girard, C., Courjon, D. and Van Labeke, D. (1990) Optical interaction between a dielectric tip and a nanometric lattice: implications for near-field microscopy, *J. Opt. Soc. Amer.* **B7**, 936-943.
9. Girard, C. and Courjon, D. (1990) Model for scanning tunnelling optical microscopy: A microscopic self-consistent approach, *Phys. Rev.* **B42**, 9340-9349.
10. Girard, C. and Bouju, X. (1991) Coupled electromagnetic modes between a corrugated surface and a thin probe tip, *J. Chem. Phys.* **95**, 2056-2064.
11. Girard, C. and Bouju, X. (1992) Self-consistent study of dynamical and polarization effects in near-field optical microscopy, *J. Opt. Soc. Amer.* **B9**, 298-305.
12. Keller, O., Xiao, M. and Bozhevolnyi, S. (1993) Configurational resonances in optical near-field microscopy: A rigorous point-dipole approach, *Surf. Sci.* **280**, 217-230.

13. Wood, D.M. and Ashcroft, N.W. (1982) Quantum size effects in the optical properties of small metallic particles, *Phys. Rev.* **B25**, 6255-6274.

14. Keller, O., Xiao, M. and S.Bozhevolnyi (1993) Optical diamagnetic polarizability of a mesoscopic metallic sphere: transverse self-field approach, *Opt. Commun.* **102**, 238-244.

15. Keller, O., Xiao, M. and Bozhevolnyi, S. (1995) Optical paramagnetic polarizability of mesoscopic particles: a study of local field corrections, *Opt. Commun.* **114**, 491-500.

16. Van Kranendonk, J. and Sipe, J.E. (1977) Foundations of the macroscopic electromagnetic theory of dielectric media, in E.Wolf (ed.), *Progress in Optics, Vol. XX* , North-Holland, Amsterdam, pp. 245-350.

# MODELLING OPTICAL RESONATORS PROBED
# BY SUBWAVELENGTH SIZED OPTICAL DETECTORS

A. Castiaux[1], Ch. Girard[2], A. Dereux[3], X. Bouju[1], J.P. Vigneron[1]

(1) Institute for Studies in Interface Sciences, FUNDP
61, Rue de Bruxelles, 5000 Namur, BELGIUM
(2) Laboratoire de Physique Moléculaire URA CNRS 772
Université de Franche-Comté
25030 Besançon Cedex, FRANCE
(3) Laboratoire de Physique Optique Submicronique URA CNRS 1796
Université de Bourgogne BP 138
21004 Dijon Cedex, FRANCE

**Abstract:** *The possibility of mapping the optical field structure inside a Fabry–Pérot resonator by using a pointed optical fiber was recently reported [1]. In this contribution, we propose a simulation of such near–field optical experiments by using a two-dimensional self-consistent model. The method based on the discretization of four different domains, i.e. the two mirrors, the glass sample and the tip, allows us a meaningful description of the evolution of the full field pattern when approaching the optical detector. In particular, this computerized work supply a direct illustration of the optical energy tranfer occurring when the tip enters the near–field zone. In this context, different tip designs are successively discussed.*

## I) Introduction

Among the various configurations used in scanning near-field optical microscopy, the Scanning Tunneling Optical Microscope (STOM) has widely demonstrated its relevance [2, 3, 4]. In this set-up, a sample is deposited on the surface of a dielectric prism illuminated in total internal reflection. The near-field, generated by the evanescent surface wave travelling along the sample, is collected by a fiber tip and transmitted to the far-field zone. The light intensity measured during the scanning process supplies local informations on both topography and index variation. Furthermore, this near-field microscopy configuration is particularly attractive because of its analogy with the well-known electron STM and also because of the relative simplicity of its principle.

In the past few years, the ongoing developement of near–field optics instrumentation stimulated a lot of theoretical analysis and several two-dimensional models were developed [5, 6, 7, 8]. These different theoretical studies clearly indicated the importance of the

95

*M. Nieto-Vesperinas and N. García (eds.), Optics at the Nanometer Scale 95–104.*
© *1996 Kluwer Academic Publishers. Printed in the Netherlands.*

96

tip–sample coupling during the scanning process. Very recently, this proximity effect was analyzed by Courjon et al. [1], and Meixner et al. [9]. They produced a standing evanescent surface wave by inserting the STOM dielectric prism between a mirror and a beamsplitter (figure 1). A Fabry-Pérot cavity is then obtained and eigenmodes are created inside the prism, which generates intensity maxima and minima on the surface at very precise positions. After collecting this surface standing wave with the usual STOM tip, they recovered in the far-field signal these interference patterns.

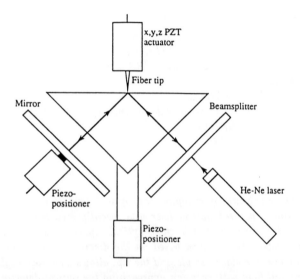

**Fig.1:** *Schematic illustration of the STOM-prism inside a Fabry–Pérot cavity*

Note that such a configuration does not only show new near-field behaviours. It also represents a direct way to control the quality–coefficient of an optical resonator. The physics of resonant cavities, particularly studied some thirty years ago to understand lasing phenomena [10], is meeting a revival of interest. New resonant systems displaying high quality coefficients (micrometric silica balls) are currently used to produce strong optical evanescent fields [11].

In a recent paper [12], we proposed a first modelling of the experimental device described in figure 1. In this contribution, we shall focus on the influence of the tip design (shape and metallic coating) on the energy conversion process. We shall also initiate a discussion about the way such resonant cavities could be used to improve the lateral resolution in STOM.

## II) A Real Space Approach for Optics in Complex Systems .

The numerical method used in the present paper is based on the Green's dyadic technique described in previous papers [13, 14]. In the past three years, different issues related to near-field optical metrology were adressed with this approach (local spectrocopy near metallic clusters, light confinement around surface defects, scattering by asymmetrical objects, etc.). These last simulations used a three-dimensional version of this algorithm. The simulations presented in this paper were achieved with a two-dimensional STOM geometry and the degenerate polarization reduces the problem to the scattering of scalar waves.

In this section, we give a brief presentation of the field propagator method that was used in our modelling. Assuming the $\exp(-i\omega t)$ time dependence, the most general vector wave equation obtained from Maxwell equations

$$- \vec{\nabla} \times \vec{\nabla} \times \vec{E}(\vec{r}) + \frac{\omega^2}{c^2} \, \epsilon(\vec{r}) \, \vec{E}(\vec{r}) = \vec{0}, \tag{1}$$

may be written as

$$- \vec{\nabla} \times \vec{\nabla} \times \vec{E}(\vec{r}) + \frac{\omega^2}{c^2} \, \epsilon_o \, \vec{E}(\vec{r}) = \mathbf{V}(\vec{r}) \, \vec{E}(\vec{r}), \tag{2}$$

where $\epsilon_o$ is the dielectric function of the external medium. In this reference medium, we know the behaviour of the electric field as the analytical solution $\vec{E}_0(\vec{r})$ of the unperturbed Helmholtz equation

$$- \vec{\nabla} \times \vec{\nabla} \times \vec{E}_0(\vec{r}) + \frac{\omega^2}{c^2} \, \epsilon_o \, \vec{E}_0(\vec{r}) = \vec{0}. \tag{3}$$

We also know the associated Green's dyadic $\mathbf{G}_0(\vec{r}, \vec{r'})$ defined by

$$- \vec{\nabla} \times \vec{\nabla} \times \mathbf{G}_0(\vec{r}, \vec{r'}) + \frac{\omega^2}{c^2} \, \epsilon_o \, \mathbf{G}_0(\vec{r}, \vec{r'}) = \mathbf{1} \, \delta(\vec{r} - \vec{r'}). \tag{4}$$

The analytical expression of $\mathbf{G}_0(\vec{r}, \vec{r'})$ in a homogeneous has the general form [15]

$$\mathbf{G}_0(\vec{r}, \vec{r'}) = \int d\vec{k} \left[ 1 - \frac{1}{q^2} \, \vec{k} \otimes \vec{k} \right] \frac{e^{i\vec{k} \cdot (\vec{r} - \vec{r'})}}{8\pi^3 (q^2 - k^2)}, \tag{5}$$

where

$$q^2 = \frac{\omega^2}{c^2} \, \epsilon_o \tag{6}$$

and $\otimes$ denotes a tensor product.

In a two-dimensional geometry, this Green's dyadic is proportional to Hankel functions, describing the circular propagation of the light around a scattering center.

The perturbation introduced in the reference medium by the presence of objects having defined geometries and compositions is entirely described by the dielectric tensor profile $\epsilon(\vec{r})$ or, if compared with the external system, by the "potential" tensor :

$$\mathbf{V}(\vec{r}) = \frac{\omega^2}{c^2} \, (1 \, \epsilon_o - \epsilon(\vec{r})). \tag{7}$$

The solution of (2) can then be obtained by solving the implicit Lippmann-Schwinger's equation

$$\vec{E}(\vec{r}) = \vec{E}_0(\vec{r}) + \int_D d\vec{r'} \, \mathbf{G}_0(\vec{r}, \vec{r'}) \, \mathbf{V}(\vec{r'}) \, \vec{E}(\vec{r'}), \tag{8}$$

where $D$ is the space domain where $\mathbf{V}(\vec{r'})$ is not zero.

Solving this equation involves a rigorous strategy:

- First, we apply an appropriate real–space discretization procedure of the space domain $D$.

- Second, we determine the field inside the perturbation from a recursive algorithm based on the parallel use of Lippmann–Schwinger and Dyson equations (see ref. [14]).

- Third, we propagate this field outside the perturbation by applying once more Lippmann–Schwinger equation.

This procedure yields a robust numerical scheme, which is able to consider large physical perturbations.

## III) Geometry of the Model

The perturbation described in figure 2, consists of two parts. The first one is the prism inside the Fabry-Pérot. It is modelled by a 2-D dielectric microprism (dielectric constant of 2.25), and two finite plates, parallel to the right angle sides. One of the plate is a gold mirror — $\epsilon$ is (-11.6,1.2) — and the second one has just a real dielectric constant of -2.0 to simulate a beamsplitter. This beamsplitter is located at a distance equivalent to half a wavelength from the dielectric prism. This distance is favourable to the creation of eigenmodes inside the microprism.

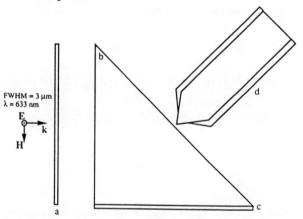

**Fig.2** : *Model of the perturbation used in our computation*

The second part of the perturbation is the probe. It is a glass fiber, etched at its apex, and covered with a thin metal coating. Here, the coating is also in gold and has a thickness of 50 nanometers. As we shall see in the next section, this coating plays an important role to confine and guide the detected near-field. During the simulation, the fiber tip is placed in contact with the prism surface. As already explained above, the method is based on a discretization of interacting objects. In this paper, all calculations are performed with a two-dimensional homogeous grid composed of squares of 50 nanometers side. Moverover, the tip scans the oblique side of the prism by steps of 70 nanometers.

The unperturbed electromagnetic wave sent on the prism through the beamsplitter is a monochromatic collimated wave. The s–polarized electric field is directed along the infinite translation direction. The incident wavelength is 633 nanometers and the FWHM of the beam is 3 microns. At the scale of our small prism, the incident wave can almost be considered as a plane wave.

## IV) Numerical Results

*-Tip-sample coupling: the role of the metallic coating*
   As a preliminary study, it is important to evaluate the role played by the coating on the energy transfer efficiency. In a previous paper [12] devoted to the same experimental configuration, we showed that a non-coated tip does not lead to an efficient lateral localization of the detection area.
   Clearly, the effective overlapping zone between the very tip and the surface can be extended enough to integrate the light energy coming from two bright fringes. For example, it can clearly be seen on figure 3 that the total intensity collected by a naked tip includes some light energy getting through its lateral sides. This effect makes difficult the discrimination between two consecutive light spots.

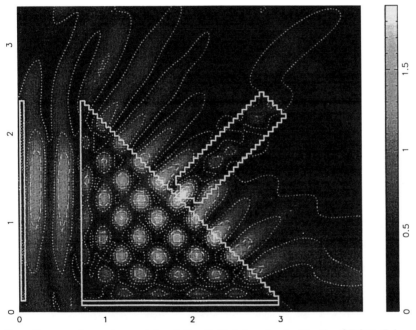

Fig. 3 : *Gray-scale contour plots of the field intensity inside the STOM Fabry-Pérot cavity described in figure 1. The uncoated pointed detector touches the surface of the prism.*

In order to get more insight on this problem, we present in figures 4, 5 and 6 three similar gray-scale representations obtained by introducing a metallic coating. Futhermore, in this context, a detailed analysis of the tip-sample interaction needs to discuss the influence of

100

the aperture extension $A_{ext}$. As expected, depending on $A_{ext}$, the tip opens a more and less efficient energy channel resulting in a radiative energy transfer above the resonator. The transfered energy rate is therefore extremely sensitive to this parameter. Moreover, the aperture must be small enough to collect information from a sufficiently localized zone.

As shown in figure 4, if the aperture is large (about 210 nm), even when detecting above a dark field zone, a high intensity coming from the neighbouring spots is collected. In figures 5 and 6, we present a second simulation by reducing the aperture extension ($A_{ext}$ = 140 nm). The shape of the tip was also modified in order to have a non-perturbating sharp tip that progressively widens to facilitate the light penetration. When comparing these patterns with figure 4, we first see that the ambiguity of lateral collection is mostly removed. The light collected by the tip in figure 5, comes essentially from the bright spot under it, and when the scanned zone is dark (as in figure 6), the light transfer from the surface to the tip is so weak that the tip core remains totally dark in our representation. Another important difference between figure 4 and figures 5 and 6 is the intervening of the tip in the resonator tune. One clearly see in figure 4 that the eigenmodes energy inside the prism is weak compared to the light energy inside the tip : the presence of the tip breaks down the resonance of the resonator and a large part of the energy is transfered to the tip. This phenomenon does not especially appear with the tip modelled in figures 5 and 6. In this case, the coupling between the tip and the resonator is lower and the energy inside the Fabry-Pérot resonator remains important.

We also considered smaller apertures and noticed, as expected, that the energy transfer to the tip becomes lower and lower when the aperture extension is reduced.

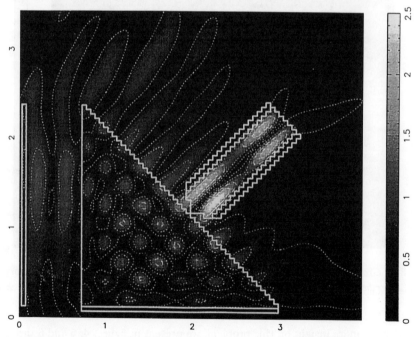

Fig. 4 : *Large aperture : mixed information coming from lateral light spots is collected by the tip*

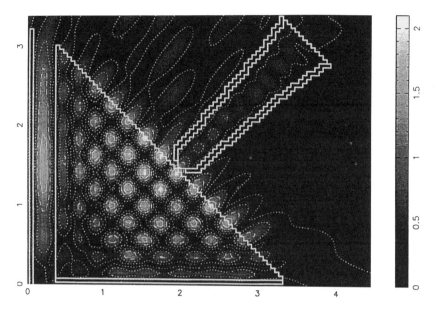

Fig. 5 : *Propagation of light inside the tip when it is located above a bright field zone*

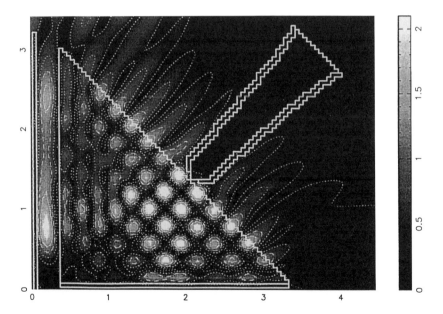

Fig. 6 : *Tip located upon a dark field zone : few light is transmitted inside the tip core*

*- Near- and far-field detection*

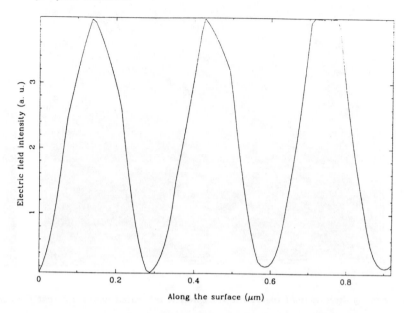

**Fig. 7** : *Shape of the surface field intensities when there is no tip above the surface*

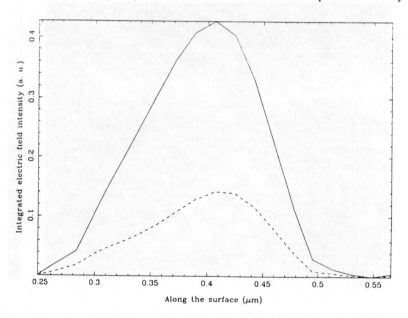

**Fig. 8** : *Integrated electric field intensity at the exit of the tip (dashed curve) and in the far-field zone (plain curve) during a scan of the surface on one period of the standing surface wave*

In the absence of the optical detector (cfr. fig.7), the field intensity on the surface present a quasi–sinusoidal shape. This standing wave results from the interferences between the two mirrors with a periodicity of about 300 nanometers. In figure 8, we present the variation of the light energy collected by tip along one period of the surface standing wave. This last simulation has been performed with the tip design already used in figures 5 and 6. On this figure, we compare the field intensity collected at the tip exit with the same quantity observed in the far-field zone, around the detection direction. Note that, for each tip–sample configuration, the far-field cross-section is integrated around a detection angle of 10 degrees. We recover the bright spot between two dark field zones, on a period of 300 nanometers. It may be seen that the dark field zones are slightly more extended than in the absence of tip (figure 7).

## V) Conclusion and discussion

In this paper, we have investigated different aspects related to the optical interaction between tips and a Fabry–Pérot resonator. Both short and long range optical couplings have been treated within the scheme of a two–dimensional self-consistent model. We have studied the energy transfer process with probes having various shapes and aperture areas. Two major effects due to the tip–sample coupling were considered. The first one was related to the extension of the coupling area. The second effect was the investigation of the transfer process itself.

Clearly, we have seen that the efficiency of the tip–surface coupling depends on the shape and the composition of the tip. As a consequence, the resonance pattern that generates the standing surface wave is more or less perturbed by the presence of the tip. The amount of light energy collected is related to the efficiency of the local channel energy as well as, to the extension of the tip–apex aperture.

The different transfer processes lead to different reproductions of the surface standing wave during a scan. When using a selective tip that localizes its interaction with the surface on a small surface portion, it is possible to recover the positions of the bright and dark zones by looking at the information transmitted by the tip. A too much perturbating tip is, for this purpose, ineffecient.

To look at small defects on the surface, we have to investigate the optimal size and shape of the tip which would improve the resolution by exploiting the resonance phenomenon. Indeed, if the interferometric system is in a resonant state, this resonance will fall down drastically when approaching the tip and could supply a direct way to enhance the resolution of scanning near-field microscopy.

## Acknowledgements

The authors especially want to thank Daniel Courjon for very helpful discussions. This work was supported by the Human Capital and Mobility project of the European Community. A. C. works under the auspices of the Belgian National Fundation for Research (F.N.R.S.).

# References

[1] D. Courjon, C. Bainier, F. Baida, Optics Comm. **110**, 7 (1994)

[2] D. Courjon, J.-M. Vigoureux, M. Spajer, K. Sarayeddine, S. Leblanc, Appl. Opt. **29**, 3734 (1990)

[3] F. de Fornel, L. Salomon, P. Adam, E. Bourillot, J. P. Goudonnet, M. Nevière, Ultramicroscopy **42-44**, 422 (1992)

[4] N. F. Van Hulst, M. H. P. Moers, B. Bögler, J. Microsc. **171**, 95 (1993)

[5] L. Novotny, D. W. Pohl, P. Regli, J. Opt. Soc. Am. A **11**, 1768 (1994)

[6] S. Bernsten, E. Bozhevolnaya, S. Bozhevolnyi, J. Opt. Soc. Am. A **10**, 878 (1993)

[7] R. Carminati, A. Madrazo, M. Nieto-Vesperinas, Opt. Comm. **111**, 26 (1994)

[8] F. Pincemin, A. Sentenac, J.-J. Greffet, J. Opt. Soc. Am. A **11**, 1117 (1994)

[9] A. J. Meixner, M. A. Bopp, G. Tarrach, Appl. Optics **33**, 7995 (1994)

[10] H. Kogelnik and T. Li, Proc. IEEE **54**, 1312 (1966)

[11] D. S. Weiss, V. Sandoghdar, J. Hare, V. Lefèvre-Seguin, J.-M. Raimond, S. Haroche, Opt. Lett. **20**, 1 (1995)

[12] A. Castiaux, A. Dereux, C. Girard, J. P. Vigneron, to be published in Ultramicroscopy, Proceedings of NFO-3

[13] C. Girard, A. Dereux, O. J. F. Martin, M. Devel, Phys. Rev. B **52**, 2889 (1995)

[14] O. J. F. Martin, C. Girard, A. Dereux, Phys. Rev. Lett. **74**, 526 (1995)

[15] H. Levine and J. Schwinger, Comm. Pure App. Math. **3**, 355 (1950)

# INSTRUMENTATION IN NEAR FIELD OPTICS

*Degree of freedom and role of polarization in near field signals*

D. COURJON, F. BAIDA, C. BAINIER AND D. VAN LABEKE
*Laboratoire d'Optique P.M.Duffieux, CNRS URA 214*
*Université de Franche-Comté, U.F.R. Sciences,*
*route de Gray, 25030 Besançon cedex, France*

**Abstract.**
    If the optical near field microscope is now considered as a new potential tool in nano-characterization, the analysis and the interpretation of the detected signals remain an unsolved problem. In this paper we will tackle the image notion first in term of degree of freedom then in term of resolution. The experimental part will be devoted to a particular case of STOM configuration in which the tip is in contact with the sample. Starting from a very simple periodic object, we will point out the crucial role of polarization on resolution and image formation.

## 1. The first attempts to break the diffraction limit in optics

The story of near field optics is tightly connected to the history of optical microscopy. We estimate that the latter appeared in the end of the sixteenth century in Holland. This first magnifying system due to a Dutch optician, Z. Jansen, was probably composed of a single lens. Some years later C. Drebbelius, a Dutch physicist, built the first compound system. From these early works, it is clear that the microscope was already similar to the modern tools we know today and that the resolution reached probably a few 10 microns or less.

    The very first attempts to increase the resolution by non conventional means (i.e. different from simple lens quality improvement) is due to G. Amici, an Italian physicist in the eighteenth century who equipped his microscope with a piece of glass set between sample and objective, in order to force the rays to converge, limiting by this way the aberration due to high angular apertures. This microscope was the ancestor of the modern immersion microscope. However the major event in modern microscopy is the development of the theory of imaging involving diffraction. This work due to E. Abbe is probably the first or one of the first evidences of the existence of a resolution limit known as *the Abbe limit*. During the first fifty years of the twentieth century we

M. Nieto-Vesperinas and N. García (eds.), Optics at the Nanometer Scale 105–117.

did not see significant progresses in imaging and microscopy concepts. New progresses started after the world war II due to the impulsion of scientists like P.M. Duffieux who proposed the use of the Fourier transform in optics [1]. The first consequence has been a dramatic simplification of the formalism leading to the introduction of new notions such as spectral response, impulse response, transfer function, etc. During the same period we note the invention of the phase contrast microscope due to Zernike in the early fifties. This event opened the door to the development of the image processing since in such microscopes, the image is the result of a non natural process leading to an output basically different from the usual perception of the object.

## 1.1. SUPERRESOLUTION BY OBJECT AND IMAGING SYSTEM BANDWIDTH MATCHING

Some years later opticians such as Lohmann, Armitage and Lukosz, proposed some techniques exploiting Fourier facilities to circumvent the Abbe limit. The basic idea emitted by Lukosz [2, 3] consists of exploiting smartly the degrees of freedom of the signal to be transmitted through the microscope. From the work of von Laue, Lukosz demonstrated that the degrees of freedom of an object (or more correctly, of the field emitted by an object) can be expressed as :

$$N_{degree} = 2N_{x,y}N_t = Cst , \tag{1}$$

where $N_{x,y}$ is the number of spatial degrees of freedom, $N_t$ is the equivalent in the time domain. The factor 2 simply expresses the existence of two polarization states. Following the papers previously cited, $N_{x,y}$ can generally be written as:

$$N_{x,y} = L_x L_y \Delta k_x \Delta k_y / 4\pi^2 , \tag{2}$$

where $L_x$ and $L_y$ correspond to the $x$ and $y$ extensions of the object. $\Delta k_x$, $\Delta k_y$ are the spatial bandwidths in the $k_x$ and $k_y$ directions respectively.

From equation 1 it appears that it is not each bandwidth in the $x$ and $y$ directions which limits the resolution, it is their product which is in fact bounded. Finally to increase the resolution in the $x$ direction it suffices to sacrifice the bandwidth in the $y$ direction. This principle is the basis of several methods developed since the sixties. It has been applied in the same period by Grimm and Lohmann [4] in a series of papers based on the object encoding by means of a suitable periodic grid. This example has been followed by a few others inspired by the synthetic aperture techniques developed in radio-astronomy. Unfortunately almost none of these techniques has known a future although they implicitly showed the way to follow in the search of superresolution.

## 1.2. SUPERRESOLUTION BY NUMERICAL RESOLUTION OF THE INVERSE PROBLEM

With the development of computer capacities appeared in the seventies the new concept of numerical inverse problem resolution. The basic idea was simple, it consisted of simulating, as faithfully as possible, the microscope working and by suitable numerical means, of retrieving the object features from experimental data. These apparently very promising techniques are unfortunately rarely mathematically stable. Moreover they need a prior knowledge about the object itself. In spite of a strong effort in theory [5, 6, 7], these methods haven't been transposed experimentally except in astronomy where it is the ultimate way to increase the resolution.

## 2. Superresolution by near field detection

As mentioned above, one of the classical ways to improve the resolution of a given information consists of sacrificing one or several degrees of freedom in order to privilege another more pertinent parameter. Following Lukosz conclusions, there is no way to exceed the classical resolution without accepting a partial loss of information in the transmitted message. However looking more closely at equation 2, we note that only $k_x$ and $k_y$ are necessary to define completely the number of degrees of freedom associated with the spectral bandwidth of the wave transmitted by the object. The role of the $k_z$ component is not explicit because it is related to $k_x$ and $k_y$ through the energy conservation law:

$$\frac{\omega}{c} = \sqrt{k_x^2 + k_y^2 + k_z^2} \tag{3}$$

Therefore a reduction of the bandwidth in the $z$ direction will not affect the number of degrees of freedom of the concerned wave. This property is constantly verified in microscopy where the spatial bandwidth in the $z$ direction is often strongly reduced. The limiting case is obviously when the $\mathbf{k}$ vector has no real $k_z$ component. This property is the basis of near field microscopy. It thus appears that the necessity of sacrificing a physical quantity to increase the precision on a more interesting parameter is not restricted to the conservation law of degrees of freedom of the transmitted message. It can be generalized to the whole parameters of the electromagnetic wave. All these considerations are not really new; they have been recently treated in the light of Heisenberg principle by Vigoureux & al [8]. As a conclusion, unlike the classical superresolution methods, near field superresolving techniques do not reduce the number of degrees of freedom of the non-directly concerned or measured quantities. They only strongly limit the domain of detection which must be as close as possible to the sample.

## 3. Meaning of the resolution notion

The notion of resolution has been clearly defined from the works of Hopkins, Duffieux and others through the introduction of the Fourier spectrum analysis. In the spectral domain, the resolution of an imaging system is simply given by the cutoff frequency of the system whatever the quality of the latter. The power of such a notion is connected to the linear transfer characterizing classical imaging transfer (in coherent, partially coherent or incoherent light as well). The attempt of proposing a meaningful definition in the near zone is not so trivial because the image is the result of the interaction of the collector/emitter and the sample itself. However a few solutions have already been proposed [9, 10].

The real problem of defining a criterion of sharpness is may be elsewhere. As it will be shown in the following, the near field structure is strongly dependent on the orientation of the electric component of the electromagnetic field. According to the relation between this one and the object structure some strong field confinements can appear in the near zone leading to spurious high resolved images. These superresolved field distributions cannot be considered as artefacts since they are the result of purely optical interactions. However they must be carefully analyzed in order to allow the retrieval of object features from detected data.

Some highly resolved images on simple structures like mica sheets have already pointed out some years ago [11]. Unfortunately these preliminary results have not been routinely obtained and remained isolated examples. More recently, theorists interested in the very short distance interaction between tip and sample showed that strong local confinements could appear on periodic or at least linear structures [12]. These results raise the question of the meaning of the detected signal. Even though some new theoretical developments show that, in certain conditions, a multi-acquisition procedure could lead to the retrieving of the exact topographic profile [13], most of the researches are oriented toward the detection of superresolved variations of the optical properties of the sample under test. Before going further a detection mode has to be defined, for unlike usual or confocal microscopy based on focusing properties of imaging system, no privilege detection plane exists. Two possibilities are offered to the experimentalist the first one consists of moving the tip in a given plane above the sample. The second one consists of using a suitable parameter like an interaction force or the signal itself to maintain the tip a controlled distance far from the surface. The first method is appreciated by theorists because the modeling is simpler, experimentally it is suitable for very flat objects (and perfectly parallel to the scanning plane). Because of no distance control the risk of artefacts is greatly reduced. However in most experimental situations (when the object is poorly identified or presents strong local topography variations), the constant height mode technique is not applicable. The distance control technique is now widely used for its simplicity of use whatever the control parameter. Its main inconvenience is the impossibility of completely decorrelating the true optical information from the control signal. Moreover it is always more noisy in comparison with the previous one.

Among the possible ways to control the distance between the tip and the sample we can distinguish 5 classes:

- distance control by optical regulation (on the evanescent field in STOM configuration [14], or on the interference in reflection SNOM [15])
- distance control by tunneling current measurement between tip and sample [16]
- distance control by shear force measurement [17]
- distance control by "tapping mode" or resonant mode force detection [18]

— distance control by contact force [19, 20].

Every method has its own advantages and drawbacks. However from experiment, it seems that the control techniques allowing a distance tip-sample as short as possible lead to the best results. As a consequence distance control based on evanescent decay or optical interference are efficient but low resolving.

## 4. Contact near field microscopy

The aim of this paper is to analyze the behavior of the near field in very precise and reproducible conditions. We have thus opted for the control by contact force. This way ensures that the tip follows the object topography rigorously without any external control signal. Its limitation is obviously the non separation between tip and sample. The microscope itself is a particular configuration based on an internal illumination (STOM or PSTM). Such a configuration first proposed in a slightly different version by van Hulst [21] has been described several times. Here we will focus our attention on the image quality, and reproducibility.

The setup is described in figure 1. As explained in ref [20] the tip is a silicon nitride hollow pyramid, which is here partially metallized. The metal covering the apex is eliminated by rubbing gently the tip on the object surface. This way was used in a slightly different procedure par U. Dürig in the early days of near field microscopy [22]. The association of the tip and a very flexible cantilever prevents the destruction of the sample or of the tip itself. Gold seems to be a good candidate because its adhesion on the silicon nitride is weak. The removal of the metal on the apex does not present any real difficulty. However it is probable that a very thin metal layer subsists.

## 5. Experimental results

The object itself is a very flat $SiO_2$ (index 1.5) grating with square grooves of 8 nm depth. The period is 383 nm. The source is a modulated polarized (5 mW power) He-Ne laser. The polarization can be rotated either by means of a simple half-wave plate. The results presented here (fig. 2,3,5) have been obtained with this technique. A better way consists of using a Pockels cell (fig. 4) which allows to verify that the evolution of the field from one polarization to another was not due to spatial shift of the light beam caused by manual handling.

We have successively imaged the grating in the two basic polarization directions i.e. in TE and TM mode assuming as mentioned further that the object plane is simply the average object plane. The incidence plane of the illumination beam is orthogonal or parallel to the grating lines. Finally in order to avoid drifts between two successive images, the polarization has been rotated while the tip is scanning the surface.

Figure 2 shows an image obtained for an illumination direction orthogonal to the object lines. We observe the high contrast of the TM image (raw data) and a frequency doubling in TE mode. Figure 3 shows a similar result in TE (a & b) and TM (c & d). Two lines have been extracted to figure out the variations from one region to another.

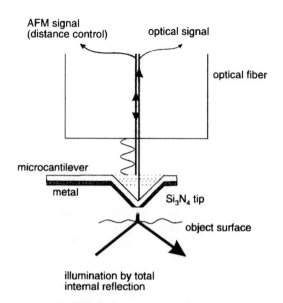

*Figure 1.* Scheme of the microscope

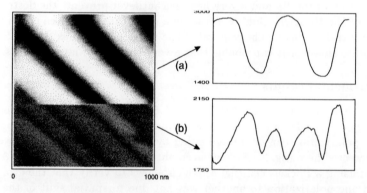

*Figure 2.* Image and cuts along one line in the TM region (a) and one in the TE region (b); the **k** vector is orthogonal to the grating lines

Again we observe the good dynamics of the TM images and the dedoubling of the TE images.

Figure 4 shows the grating image when the **k** vector is parallel to the grating lines. The polarization state goes from TE to TE passing by TM. We do not observe any fringe dedoubling; TE and TM mode images have similar aspect.

The last example is presented in figure 5 where a suitable polarization mixture can lead to "good images"; the notion of quality being questionable.

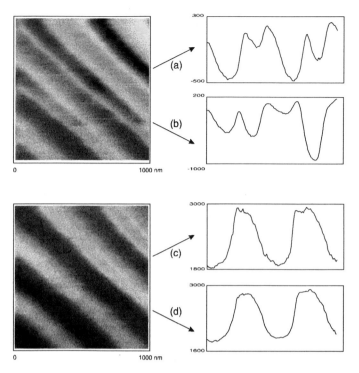

*Figure 3.* Image and cuts in two regions, in TE (a & b) and in TM (c & d) modes; the **k** vector is orthogonal to the grating lines

## 6. Modeling of the interaction

At the moment three methods are mainly used for modeling the interaction between the light and the sample in the near zone. One is known as MMP method (multiple multipole method) [23], the second one is called Green function or propagator function method [12]. Finally the last one is derived from diffraction theory and often based on perturbative approaches (resolution of Maxwell equations) [10, 24, 25, 26]. The proposed technique belongs to the last family. This method can take into account the possible coupling between tip and sample by applying a matrix formalism [27]. Moreover it can simulate realistic experimental conditions such as metal coated or uncoated tips with respect to any polarization state. The main drawback is the necessity to assume Rayleigh conditions (continuity of fields inside the structures). However the type of object under test makes this assumption perfectly valid. A second drawback concerns the infinite periodicity and thus the infinite lateral extension of both sample and tip. This problem is due to the fact that the calculation is carried out in the Fourier space.

It is clear that, strictly speaking, the contact mode cannot be simulated with our model. However it has been approached by assuming that the tip follows the topography

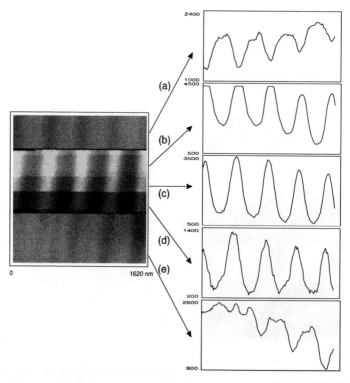

*Figure 4.* Image and cuts in five regions, from TE (a) to TE (e) mode passing by 1 m (nearly (c)); the **k** vector is parallel to the grating lines

8 nanometers far from the object surface. In order to ensure realistic conditions, the actual tip displacement has been simulated by convolving the object profile and a suitable rectangle function introducing a small smoothing of the scanning curve as shown in figure 6. This function is nothing but the aperture of the tip itself.

In the simulation the aperture angle corresponds to the actual shape of a $Si_3N_4$ AFM tip (70 ° of aperture angle). The optical index is 2.1. Although the model seems to be far from the exact tip shape because of the infinite metal slab, both experiments and other more sophisticated nano-emitter models have shown that only the discontinuity due to the apex of the cone effectively participates in the detection of the evanescent light. The presence of the metal slab does not play any significant role in the light detection process.

The two basic polarizations have been studied for orthogonal and parallel illumination. The notion of polarization state is somewhat ambiguous because it is defined in relation with a plane of incidence which contains the light rays and the normal to

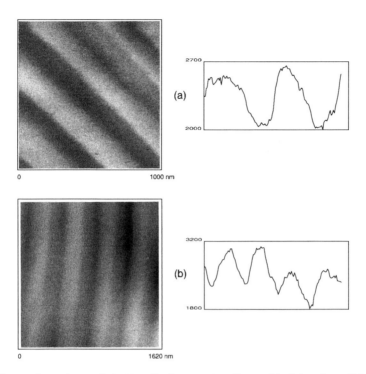

*Figure 5.* Images for a given polarization, the **k** vector is orthogonal in (a) and parallel in (b) to the grating lines

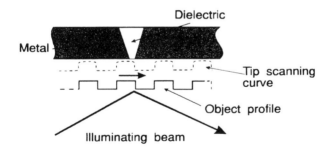

*Figure 6.* Schematic diagram of the setup used for the modeling

the object surface. Therefore such a plane depends on the object topography and will change according to the considered point in the sample. For the sake of simplicity the polarization is always defined according to the average surface of the sample. Therefore

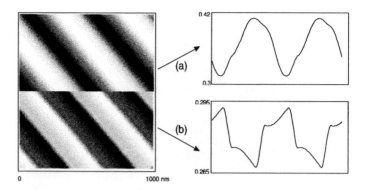

*Figure 7.* Simulation in TM (a) and TE (b) modes. The **k** vector is orthogonal to the grating lines

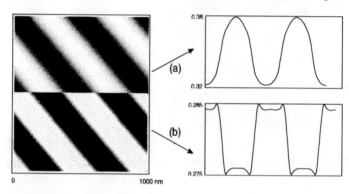

*Figure 8.* Simulation in TM (a) and TE (b) modes. The **k** vector is parallel to the grating lines

it has no real physical meaning. It seems better to define the relation between sample structures and electric field direction.

Figure 7 shows the simulated intensity in TM (a) and TE (b) modes. The **k** vector is orthogonal to the grating lines. We note:

- the asymmetry of the obtained profiles (well visible in TE mode). This effect is due to the asymmetry of the illumination beam.
- the higher contrast in TM than in TE mode.
- the quasi-absence of ripples and undulations on the cut in TM mode. However the walls seem to be better resolved in TE than in TM. These results are in good agreement with our experimental data (see fig. 3) .

In figure 8 the **k** vector is parallel to the grating lines. The main remark concerns the perfect symmetry of the grating line profiles. About contrast and ripples as in orthogonal illumination, the contrast is weaker in TE mode which always contains more ripples.

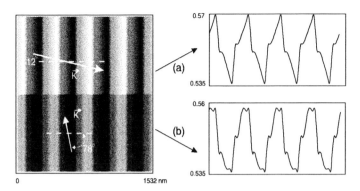

*Figure 9.* Effect of a small rotation of the plane of incidence versus grating lines; 12 degrees in (a) and 78 degrees in (b), both in TE mode

An interesting effect appears in figure 9. When the plane of incidence is not perfectly parallel to the grating lines, (here rotation of 12°) the line profile dramatically changes and loses its symmetry. This sensibility to the $\vec{E}$ and $\vec{B}$ against the object structure explains the difference between the images obtained experimentally. This effect is less visible in orthogonal illumination but it is still perceptible as shown in figure 9 (a).

## 7. Discussion of experimental results in the light of theoretical modelings

The conclusion concerning our results is first that the detected signals are actual optical ones. They are also reproducible and point out the fundamental role of the polarization in relation with the object structure. More precisely in this case of metallized tips we note that the TE polarization mode seems to be the ideal configuration for interacting with the object (assuming orthogonal illumination direction versus grating lines). When the sample presents some linear features the interaction effect can be dramatically enhanced on the object rims (see fig. 2 for instance). The TM mode seems to be more adapted to microscopy. We note the good contrast of the image and the absence of parasitic ripples. These observations are in agreement with the theoretical modelings presented in reference [24] and [25] and in this communication. In these works, it appears that the behavior of metallic tips differs fundamentally from that of dielectric tips. The difference is mainly due to the existence of a non negligible magnetic dipole in the case of metallic apertures or metallic tips. This magnetic dipole induces effects which can significantly modify the field structure. However if the case of dielectric tip is rather clear (because there is no magnetic effect), in the case of metallic tips both electric and magnetic susceptibilities intervene in the detection process. Anyway whatever the nature of the tip it appears that the good contrast observed is probably due to the periodicity of the grating interacting strongly with the light field (whose period is not very different). This interaction is nothing but a resonant effect enhancing artificially the line contrast. The sensibility to line orientation shown in figure 9 confirms this feeling. Some previous results obtained in constant distance mode and with a tapered uncoated fiber confirm our analysis [14].

## 8. Conclusion

Near field optics can lead to superresolved images without any reduction of the information present in the object itself. In this sense it is probably the most powerful tool available in the nano-characterization field. However it is clear that the recorded field intensity variations are the result of a very complex interaction between tip, sample and incident field. A common point between experimental results and theoretical modeling is the necessity to bring the nano-collector or the nano-emitter as close as possible to the sample itself. In this sense the contact near field microscopy seems to be an ideal case even if the non destructive and the non intrusive character often claimed in optical analysis is questionable. From both theoretical and experimental results, it appears that the detected signal strongly depends on the object itself. Object types such as gratings seem to be bad candidates for testing the capability of near field microscopes to provide significant information. We have more particularly shown that the contrast can be greatly enhanced by the propagation in a periodic structure. Some previous works using the same theoretical models have already pointed out this particularity. If this strong object dependence can be considered as a real drawback in microscopy, it could be turned into advantage in nano-optics for example for generating optical resonators in the nanometer range. Finally, the presence of metal on the very tip is probably the way to follow because of simultaneous existence of electric and magnetic dipoles able to couple with the object near field.

## Acknowledgements

The authors are indebted to Christian Girard for numerous and fruitful discussions about polarization and confinement in the near zone.

## References

1. P. M. Duffieux. *L'Intégrale de Fourier et ses Applications à l'Optique*. Masson, 1970.
2. W. Lukosz. Optical systems with resolving powers exceeding the classical limit. *J. Opt. Soc. Am.*, 56, No 11:1463–1472, November 1966.
3. W. Lukosz. Optical systems with resolving powers exceeding the classical limit.II. *J. Opt. Soc. Am.*, 57, No 7:932–941, July 1967.
4. M. A. Grimm and A. W. Lohmann. Superresolution image for one-dimensional objects. *J. Opt. Soc. Am.*, 56(9):1151–1156, September 1966.
5. M. Bertero, G. A. Viano, and C. de Mol. Resolution beyond the diffraction limit for regularized object restoration. *Optica Acta*, 27(3):307–320, 1980.
6. A. Lannes, S. Roques, and M. J. Casanove. Stabilized reconstruction in signal and image processing 1.Partial deconvolution and spectral extrapolation with limited field. *Journal of Modern Optics*, 34(2):161–226, 1987.
7. E. Lantz and J. Duvernoy. Stability of model and selection of parameters : application to metrology in optical microscopy. *Journal of Modern Optics*, 36(9):1213–1226, 1989.

8. J. M. Vigoureux and D. Courjon. Detection of non-radiative fields in light of the Heisenberg uncertainty principle and the Rayleigh criterion. *Applied Optics*, 31(16):3170–3177, June 1992.

9. C. Pieralli. Statistical estimation of point spread function applied to scanning near-field optical microscopy. *Optics Commun.*, 108:203–208, June 1994.

10. Jean-Jacques Greffet, Anne Sentenac, and Rémi Carminati. Surface profile reconstruction using near-field data. *Optics Commun.*, 116:20–24, 1995.

11. D. Courjon, J. M. Vigoureux, M. Spajer, K. Saraeyddine, and S. Leblanc. External and internal reflection near field microscopy: experiments and results. *Appl. Optics*, 29:3734–3740, 1990.

12. C. Girard and A. Dereux. Optical spectroscopy of a surface at the nanoscale: A theoretical study in real space. *Phys. Rev. B*, 49 no 16:11344–11350, 1994.

13. N. García. Near field optics imaging. In *NATO ARW on Near Field Optics*. Miraflores, September 1995.

14. D. Courjon, C. Bainier, and M. Spajer. Imaging of submicron index variations by scanning optical tunneling. *J. Vac. Sci. Technol.*, 10(B):2436–2439, 1992.

15. M. Spajer and A. Jalocha. The reflection near field optical microscope: an alternative to STOM. In D. W. Pohl and D. Courjon, editors, *Near Field Optics*, volume 242 of *NATO ASI Series*, pages 87–96. Arc et Senans, France, Kluwer Academic Publishers, 1993.

16. M. García-Parajo, E. Cambril, and Y. Chen. Simultaneous scanning tunneling microscope and collection mode scanning near-field optical microscope using gold coated optical fiber probes. *Appl. Phys. Lett.*, 65(12):1498–1500, September 1994.

17. E. Betzig, P. L. Finn, and J. S. Weiner. Combined shear force and near-field scanning optical microscopy. *Appl. Phys. Lett.*, 60(20):2484–2486, May 1992.

18. R. Bachelot, P. Gleyzes, and A. C. Boccara. Near-field optical microscope based on local perturbation of a diffraction spot. *Optics Letters*, 20(18):1924–1926, September 1995.

19. N. F. Van Hulst, M. H. P. Moers, and B. Bölger. Near-field optical microscopy in transmission and reflection modes in combination with force microscopy. *Journal of Microscopy*, 171(Pt2):95–105, August 1993.

20. F. Baida, D. Courjon, and G. Tribillon. Combination of a fiber and a silicon nitride tip as a bifunctional detector; first results and perspectives. In D. W. Pohl and D. Courjon, editors, *Near Field Optics*, volume 242 of *NATO ASI Series*, pages 71–78. Arc et Senans, France, Kluwer Academic Publishers, 1993.

21. N. Van Hulst, F. B. Segering, and B. Bölger. High resolution imaging of dielectric surfaces with an evanescent field optical microscope. *Optics Commun.*, 87:212–218, 1992.

22. U. Dürig, D. W. Pohl, and F. Rohner. Near-field optical-scanning microscopy. *J. Appl. Phys.*, 59(10):3318–3327, May 1986.

23. L. Novotny, C. Hafner, and D. W. Pohl. The multiple multipole method in near-field optics. In *NFO-3*, volume 8, pages 31–32. Brno, EOS Topical Meeting, 1995.

24. D. Barchiesi and D. Van Labeke. Scanning tunneling optical microscopy (STOM): Theoretical study of polarization effects with two models of tip. In D. W. Pohl and D. Courjon, editors, *Near Field Optics*, volume 242 of *NATO ASI Series*, pages 179–188. Arc et Senans, France, Kluwer Academic Publishers, 1993.

25. D. Van Labeke and D. Barchiesi. Probes for scanning tunneling optical microscopy : a theoretical comparison. *J. Opt. Soc. Am. A*, 10(10):2193–2201, October 1993.

26. M. Nieto-Vesperinas and A. Madrazo. Light scattering by tips in front of surfaces. theory for scanning photon tunneling microscope. In *NATO ARW on Near Field Optics*. Miraflores, September 1995.

27. F. Baida. Microscopie hybride : association d'un microscope optique en champ proche et d'un microscope à forces atomiques. Principe et réalisation. Thèse de doctorat de l'Université de Franche-Comté, Besançon. Fev.1995.

# Effect of the Coherence in Near Field Microscopy

F. de Fornel, L.Salomon, J.C. Weeber, A. Rahmani, C. Pic, A. Dazzi

*Equipe Optique Submicronique*

Laboratoire de Physique URA 1796

Faulté des Sciences Mirande BP138

21004 DIJON cedex    FRANCE

## I Introduction

The observation of samples with optical technics reveals the effect of the degree of coherence of the light source. As reported by Goodmann, an edge appears as surrounded by some oscillations if the source is highly coherent [1]. The shape of the image of the edge is linked to the diffraction pattern and depends on the coherence of the source [2]

The resolution depends of course on the degree of coherence of the source. With a simple lens of diameter D and at the focal distance z, two points can be distinguished if they are separated by a distance h: with a coherent source $h_c=1.64\lambda z/D$ and with an incoherent one $h_i=1.22\lambda z/D$, [2].

The use of incoherent source reduces the effect of the interference on the image formation. The effect of the degree of coherence of the source has been observed in near field microscopy with a SNOM, by E. Betzig and with the PSTM by N. van Hulst or T. Ferrell [3-5].

The results discussed in this paper concern only the effect of the coherence on the formation of the images obtained with a PSTM.

## II Effect of the degree of coherence on the image formation of calibrated samples.

### a) Introduction

We have chosen to show the results obtained for two sorts of samples. The first one is a flat sample with a very low roughness. The second one presents a geometry as simple as a step or a rectangle.

*M. Nieto-Vesperinas and N. Garcia (eds.), Optics at the Nanometer Scale 119–130.*

The results obtained with the first sample will allow us to introduce the concept of speckle in near field. The images taken with the second sample will be compared with numerical simulations.

### b) Speckle in near field

Figure 1 is an image of a flat sample made of glass which exhibits a low roughness; the angle of incidence of the incoming light is 60°. The source is a He Ne laser. A few remarks can be expressed about the image. The image looks flat and a line profile over a distance of 30μm shows a variation of the apparent height of the order of few nanometers, which partly correspond to the noise of the experimental setup and to the curvature of the piezoelectric tube used for the scan.

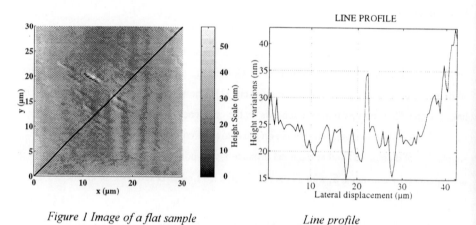

*Figure 1 Image of a flat sample*                    *Line profile*

This image presents a periodical modulation which is perpendicular to the plane of incidence of the illuminating light. These oscillations can be created by the interferences existing between the different waves reflected by the lateral edge of the sample or by the reflection of the light at the different interfaces. Such interferences were described by Salomon and Mexnier [6,7]. The interference pattern depends on the wavelength [6].

After a few scans and after changing the lateral position of the tip, we obtained the following image on which three particles can be observed

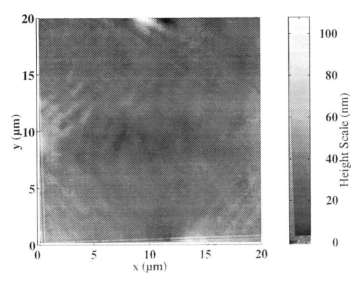

*Figure 2 Image of a flat sample with a few particules on it*

Every particle diffracts the light and creates a pattern as described by van Hulst [4]. The light diffused by the particles can interfere and creates the pattern observed on figures 2 and 3. Interference pattern is visible from a distance to the particles farther than 10μm.

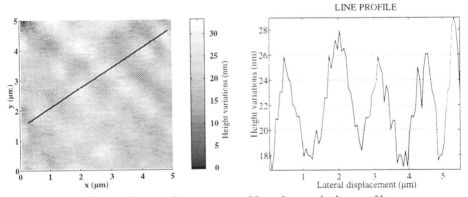

*Figure 3 Image of a scan range of 5μmx5μm and a line profile*

When comparing with the image figure 2, this image appears totally different, within these conditions, it is impossible to identify the structure of the substrate. If a white source is put in place of the He Ne laser, the image is totally different, (figure 4). The interference pattern have disappeared. The shape of the surface looks more actual. The line profile made on a zoom of the image fig.4 shows details as small as 100nm. Besides, the image of the surface seems very flat.

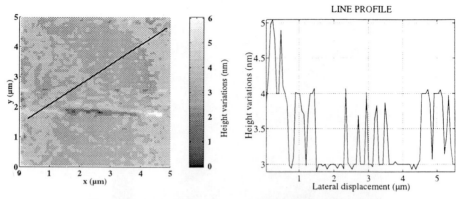

*Figure 4a Image of the same range taken with a white source*

Doing a scan on 30µmx30µm, the same range than for the figure 1, gives a image, fig.5, on which the particules are better localized.

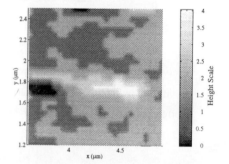

*Figure 4b Zoom of the figure 4a*

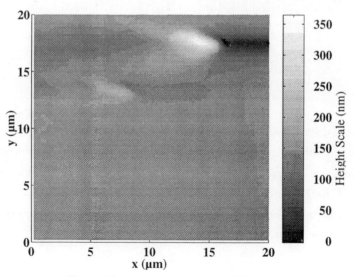

*Figure 5 Image of a scan range of 20µmx20µm.*

When the centers of diffusion are in limited number it is possible to recognize the interference pattern. But the roughness of the sample is often described by a random function. When the sample is illuminated with a coherent light source, the random distribution creates a pattern called speckle. Using incoherent source reduces the speckle as it is shown on figure 6. These results have been confirmed numerically by Greffet and Carminati [7].

### c- Effect of the coherence on the resolution power in near field

We have yet observed the effect of the coherence on the image formation of simple sample such as a single step on a quartz substrate, figure 6 [8]:

*Figure 6 Images of a step ( a: He Ne laser and b: white source)*

We used different kinds of sources with the corresponding length of coherence Lc: a He-Ne laser (Lc is a few meters), a He-Cd (Lc is a few tens of centimeters), a monochromatic source (Lc=100µm) and a white source (4µm< $\lambda$ <8µm) (Lc=1µm). We have observed that for small values of the coherence length the image looked closer to the actual structure. These results are confirmed by our simulations, figure 7. The differential method has been used to model the field above the sample [10,11]. The results reported on figure 7 confirms the experiments.

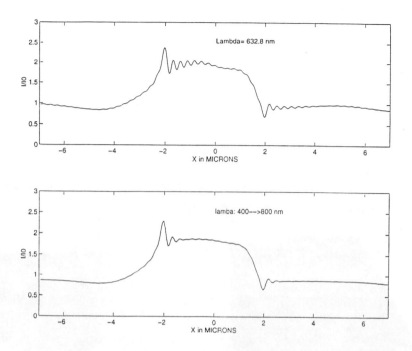

*Figure 7 Effect of the coherence on the image of a rectangle on a flat substrate in quartz*
*h=60nm,w=4μm.*

In order to go further into the analysis of the effect of the coherence, we have simulated a sample with two identical structures separated by a distance d, see figure 8.The aim of this simulation is to determine the effect of the coherence on the resolution. The geometry of the sample is described by the figure 8.

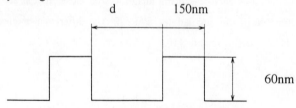

*Figure 8 Geometry of the simulated sample.*

The figure 9 describes the intensity of the field near the sample. We give a few examples for the TE and TM polarizations.

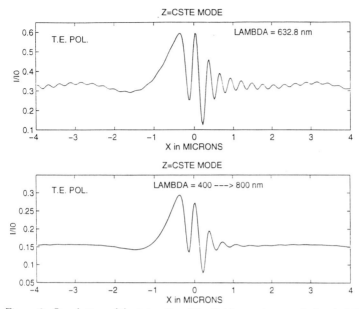

*Figure 9a Simulation of the intensity of the field near the sample for d=350nm*

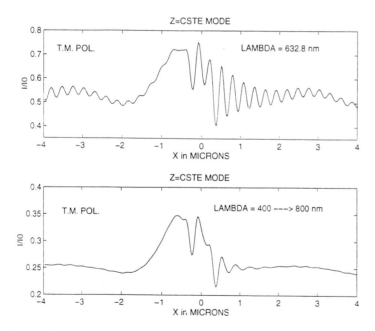

*Figure 9b Simulation of the intensity of the field near the sample for d=600nm*

126

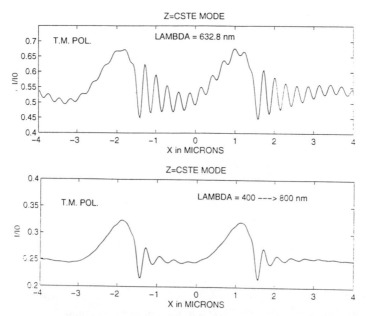

*Figure 9c Simulation of the intensity of the field near the sample for d=3μm*

This results show that, as in classical microscopy, the resolution power is improved by the use of sources with a low degree of coherence. The degree of coherence is related to the numerical aperture as well as to the spectral width of the source. Experimental results are in progress but the first images taken with a laser confirm the impossibility to distinguish two objects separated by a distance of 500 μm at λ=0.633μm.

*Figure 10 Image of a double rectangle sample*

Before the conclusion, the last part is devoted to the presentation of images which confirm the interest to use an incoherent illumination.

## III SAMPLES ILLUMINATED WITH COHERENT AND INCOHERENT SOURCES

### a) Observation of a dopped lithium niobate slide

**lithium niobate slide**

The sample is described by the following schematic. The variation of the refractive index is obtained after the implantation of ions ($H^+$). The energy of the protons flux was equal to $10^{16}$ ions/cm$^2$ The refractive index of the doped part is complex. The sample has an increase of the absorption where the implantation has occured.

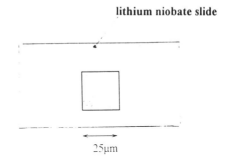

25µm

*Fig 11 Schematic of the implantation*

The implantation is made through a mask with squared holes. The images, where the sample is doped, present a hole.The depth is of the order of 20nm, it is due essentially to the variation of the refractive index. Indeed an image taken with an AFM (Atomic Force Microscopy) presents a surface with a hole of only 5nm.

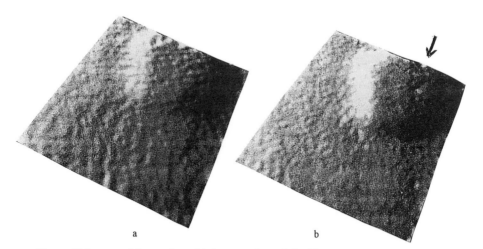

a                                    b

*Figure 12 Images of the implanted lithium niobate slide. The source is **a**-a He-Ne laser,*
*b- a incoherent monochromatic source.*

Images have been taken with the He Ne laser, a He-Cd laser and an incoherent source (λ=0.633μm). We can notice that the small detail indicated by an arrow is visible only when using the He Cd laser or the incoherent monochromatic source.

## b)Observation of fluorescent polystyren spheres.

The images of the fluorescent spheres have been obtained by collecting all the light in the near field of the spheres. In other words at the wavelength at the maximum of the excitation, the tip fiber collects the light at this wavelength as well at the wavelength corresponding to the emission.

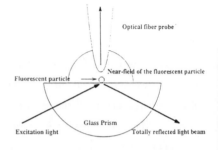

*Figure 13 Schematic of the measurement*

Because of the geometry of the sphere and the roughness of the substrate it is impossible to obtain any clear image by using a coherent source like a He Ne laser, but the use of incoherent source allows to image the sample.It has been possible to localize spheres of 50nm and 970nm of diameter. The wavelength is varying on a range centered on the maximum of the excitation of the fluorescence. The set of images shows that for a peculiar wavelength the shape of the spheres looks more confined.

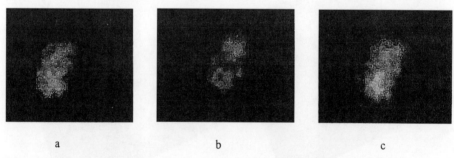

a                                     b                                     c

*Figure 14 Images of fluorescent polystyren spheres obtained at different wavelengths:*
*a- λ = 612 nm, b- λ = 535 nm, c- λ = 532 nm.*

# IV CONCLUSION

The results described in this paper show how the coherence of the source plays a role in image formation in near field microscopy. The speckle observed in far field exists also in near field and is linked to the roughness of the sample. The resolution depends on the coherence and is improved by illuminating the sample with incoherent source. It has been shown with two examples that incoherent illumination can be the only way to obtain usable images of certain samples (for sample with a well defined geometry like AgBr grains[12], implanted Niobiate Lithium or polystyren spheres). However we have to keep in mind that the interference pattern contains information which can be very helpful and even necessary to the analysis of samples

ACKNOWLEDGMENTS

The authors thank C. Amra from the Ecole de Physique de Marseille (Marseille France) and H. Launois from the L2M (Bagneux France) for providing the samples.This work was performed within the CNRS ULTIMATECH program.

REFERENCES

[1] Goodmann J.W. "Introduction à l'optique de Fourier et à l'holographie", Masson, Paris(1972).

[2] Ash E.: "Scanned image microscopy" , Academic Press, London (1980).

[3] Betzig E.: PhD Thesis, Cornell University (1986)

[4] van Hulst N., Segerink F.B., Bölger B.: Opt. Comm. 87, 214, (1992).

[5] Ferrell T.L., Goudonnet J.P., Reddick R.C., Sharp S.L.: J. Vac. Scien. Technol. B9,2, 525-530 (1991).

[6] Salomon L. Thesis , Dijon, France (1991)

[7] Meixner A., Bopp M., Tarrach G.: Applied Optics, 33, 34, 7995-8000 (1994)

[8] Greffet J.J. and Carminati R: Proc. Near Field Optics and Related Technics, NFO3, BRNO 9-11 May 1995.

[9] de Fornel F., Salomon L., Adam P., Goudonnet J.P. and Guerin Ph.: Optics Letters, 19, 14, 1-3 (1994).

[10] Vincent P.: Electromagnetic Theory of Grating, ed.R.Petit (Springer Verlag, 1980) pp101-121.

[11] Weeber J.C., de Fornel F. and Goudonnet J.P. submitted for publication.

[12] Pic C., de Fornel F., Weeber J.C., Lesniewska E., Martin D.,Guilement J., Goudonnet J.P.: Proc. Near Field Optics and Related Technics, NFO3, BRNO 9-11 May 1995.

# SCANNING INTERFEROMETRIC APERTURELESS MICROSCOPY AT 10 ANGSTROM RESOLUTION

H.K. WICKRAMASINGHE, Y. MARTIN and F. ZENHAUSERN
*IBM Research Division, T.J.Watson Research Center,*
*P.O.Box 218, Yorktown Heights,*
*New York, NY 10598*

## 1. Introduction

Near-field scanning microscopy at microwave frequencies (1) and its extension to the visible region (NSOM)(2, 3) at around 50 nm resolution has attracted much attention (4-7). Recently, a Scanning Interferometric Apertureless Microscope (SIAM) was introduced (8, 9, 14) where the scattered electric field variation due to a vibrating and scanning probe tip in close proximity to a sample surface is measured by encoding it as a modulation in the phase of one arm of an interferometer. Here, we review the SIAM technique and present images of various samples at a resolution of 1 nm - almost two orders of magnitude superior to other NSOM's. A basic theory based on coupled dipoles is put forward and compared with experiments. It shows that the contrast mechanisms are fundamentally different to those in regular near-field optical microscopes due to a unique dipole-dipole mechanism. Furthermore, the theory predicts the ability to measure complex susceptibility down to the atomic level.

## 2. Principle of Operation

Conventional near-field optical microscopes based on apertures - typically drawn and coated single mode optical fibers are limited in resolution by the optical penetration depth of the metal coatings used to confine the light within the fiber to a resolution around 50 nm. Another difficulty with these systems is that in the final taper region of the fiber -where the guided mode is cut-off, there is typically a loss of six orders of magnitude in going down to a 50 nm aperture size. In SIAM, we introduce a technique whereby the 50 nm barrier can be clearly surpassed while at the same time not losing the six orders of magnitude in optical power illuminating imaging region. The key steps of this new microscope are:

- Use of a microscope objective to tightly focus a laser to a diffraction limited spot on the sample

*M. Nieto-Vesperinas and N. García (eds.), Optics at the Nanometer Scale* 131–141.
© *1996 Kluwer Academic Publishers. Printed in the Netherlands.*

- Use of a very sharp tip to "sample" a tiny region of this spot, generating a scattered spherical wave

- Adding a concentric reference wave to the scattered wave

- Measuring the amplitude or phase of the resulting electric field of the combined wave in a very sensitive, photon noise limited interferometer

- The loss part of the field interaction with the sample is proportional to the amplitude change in the measured field and the dispersive part of the interaction is proportional to the phase change in the measured field

- Vibrating the tip very close (within 1 nm) to the sample so as to exploit the rapid change in the scattered signal as a function of tip-feature distance (due to the strong dipole-dipole interactive tip-feature coupling) ;this helps localize the measurement region to approximately a tip diameter

- Detecting the interferometric signal at the tip vibration frequency in a lock-in amplifier and recording this in a computer as the sample is raster scanned relative to the tip to form an image

## 3. Experimental

Experiments were performed in the transmission mode (9). An incident laser beam is focused on the back surface of a transparent substrate holding the sample (Fig.1). A z-vibrating tip ($f_z$ = 250 kHz, spring constant = 20 N/m, tip-end diameter $\simeq$ 5 nm, vibration amplitude = 6-10 nm) is approached to the focused spot and stabilized 1 to 2 nm over the sample surface using an attractive-mode AFM (11). The return beam $E'_r + E_s$ ( reflection from substrate plus tip-sample scattering) is detected using an interferometer, by combining it with a reference beam, $E_r$. The output signal of the interferometer measures either the amplitude of ($E'_r + E_s$) or its phase difference with $E_r$ which represents the contrast mechanisms.

The experimental data in Fig.2 show an AFM (a) and simultaneously recorded optical image (b) of a cleaved mica surface. The AFM image shows a 0.25nm monoatomic terrace on the left which appears as a bright scattering region in the optical image. The AFM also shows a step in the center comprising approximately 15 atomic layers. The optical image shows the same feature as a bright band across its center but in addition shows much sharper detail (see arrows) not visible in the AFM image. These features could be sub-surface defects in the mica as they do not appear in the topography; the smallest discernible feature size visible optically is approximately 1 nm and corresponds to an interferometric phase shift of $2.10^{-5}$ radian.

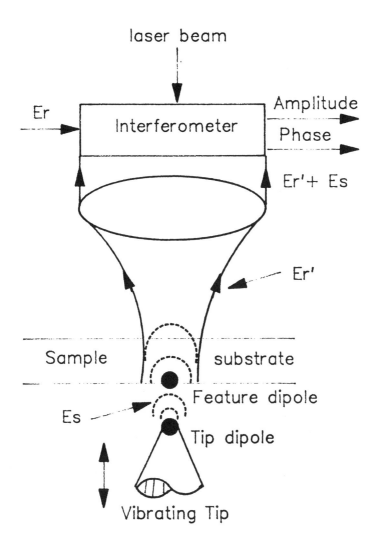

Fig.1 Principle of SIAM method.

134

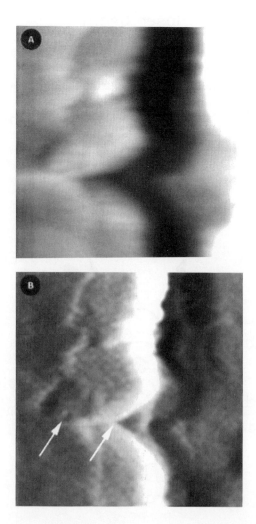

Fig.2  a) Attractive mode AFM image showing atomic terraces on the mica
      b) Simultaneously recorded SIAM image revealing bright
         scattering regions corresponding to the atomic terraces but
         also showing fine sub-surface features not present in the AFM
         image; smallest resolvable feature $\simeq 1$ nm.

In the next experiment, (Fig.3) index matching oil was dispersed into tiny droplets on cleaved mica by repeatedly scanning an AFM tip across its surface (for approximately 1 hour at a force gradient of 0.1 N/m). The data were acquired at a force gradient an order of magnitude lower and both AFM and optical images were simultaneously recorded. The AFM topography image - which represent constant force gradient contours - show the oil droplets as bumps on the mica surface. Certain regions also show dips (top right arrow). These indicate that some oil droplets are probably charged and therefore locally reduce the overall force gradient on the tip due to electrostatic interactions. The optical image (3b) clearly shows the oil droplets as enhanced scattering centers; the dips in the AFM image showing up as optically bright regions with no sensitivity to charge. Note that a feature in (3a) (top left arrow) is totally absent in the optical image. One can characterize this topographic feature producing little scattering and showing optical properties quite different to those of the oil droplets. The smallest feature resolved optically (bottom right arrow in (3a) ) is approximately 1nm as in the earlier experiments and a phase change of $10^{-4}$ radian is measured.

Experiments have indicated that in some situations the image contrast can switch depending on the optical parameters at the tip-end - from high contrast to low contrast or inverted contrast. The test sample of figure 4 was prepared by spinning 45 nm fluorescent polystyrene spheres (10) on a cleaved mica surface. Figure 4(a) shows the AFM image of an island of polystyrene spheres - the topography showing individual spheres. Figure 4(b) shows the simultaneously recorded optical image, with interferometric phase modulation of about $10^{-3}$ radian. A sudden change clearly appears in the optical image, whereas the AFM image shows virtually no change (on careful examination of AFM line scans, a minute change in the tip z-position of approximately 5 nm is detected). This can be explained if the tip picked up a particle as it raster scanned the sample, perhaps a fluorescent dye molecule, which changed the phase of the scattered field. In the top region of Fig.4b, the spheres appear as a low signal in a bright background; here the scattered electric field $E_s$ is expected to have a component $\frac{\pi}{2}$ out of phase with $E'_r$. In the lower region, the contrast is weaker but reversed, and implies that $E_s$ has a component $\pi$ out of phase as compared with the situation in region 1. Several similar results demonstrate that small particles (around 5 nm) can significantly affect the scattered field, and reveal the high sensitivity of the SIAM technique to small changes in the tip-sample optical interaction. As a further evidence for the source of contrast, a 100 nm chromium grating was scanned at constant height with the AFM in order to remove any tip-height related signal. While no detail was visible in the AFM image, the optical image showed strong contrast due to dipole-dipole coupling, at spacings of about 10 nm or less.

136

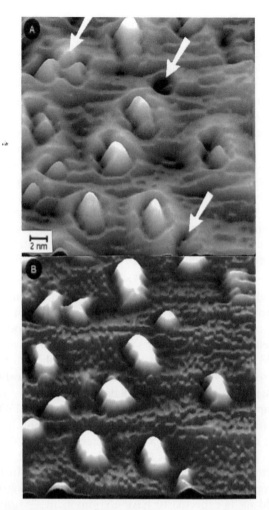

Fig.3 Dispersed Oil Droplets on Mica.
    a) Attractive mode AFM topography image showing most droplets
    as bumps in the image while some are shown as dips due to
    electrostatic charging effects.
    b) Simultaneously recorded optical image showing the oil
    droplets as bright scattering regions. Some features in
    the AFM image are absent in the optical image and others
    are inverted in contrast (arrows) - see text;
    smallest resolvable feature $\simeq$ 1nm.

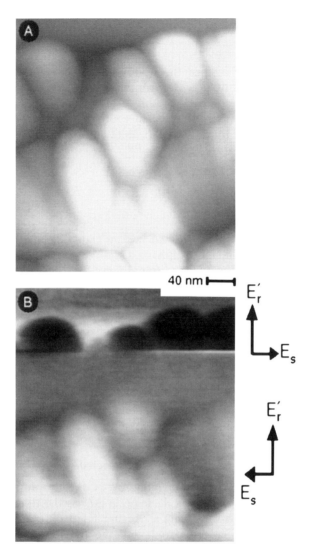

Fig.4    Images of 45 nm Fluorescent Polystyrene Spheres on Mica
a) Attractive mode AFM topography
b) Simultaneously recorded SIAM image showing contrast
reversals in different regions caused by a small
particle ( $\simeq$ 5 nm) sticking to the tip.

## 4. Theory

A theory based on coupled dipoles can be derived to understand the contrast mechanism in SIAM. Since the optical dipole interaction varies as $r^{-3}$, the measured signal primarily derives from the tip end. One can therefore model tip and sample-feature as spheres of radius a (Fig.1), with polarizabilities $\alpha_t$ and $\alpha_f$ and tip-feature spacing r, immersed in a driving electric field $E_i$. The general equation (12) that describes the polarizability modulation $\Delta\alpha$ due to the coupling between the tip and feature is:

$$\Delta\alpha = \frac{2\alpha_t\alpha_f}{(r^2 + a^2)^{\frac{3}{2}}} \tag{1}$$

The modulation $\Delta E_s$ of the scattered field $E_s$ caused by the polarizability modulation $\Delta\alpha$ can be calculated by applying a scattering matrix (S) treatment for small particles (13). The spherically scattered wave at a distance d in the far field is given by

$$E_s = \frac{E_i}{ikd}(S) \quad ; \quad \Delta E_s = \frac{E_i}{ikd}(\Delta S) \tag{2}$$

For particles significantly smaller than 50 nm, where $ka \ll 1$ ($k = 2\pi\frac{n}{\lambda}$ being the optical propagation constant in a medium of refractive index n), the relevant scattering matrix component S (which has both real and imaginary components) can be written in terms of the polarizability $\alpha$.

$$S = ik^3\alpha \quad ; \quad \Delta S = ik^3\Delta\alpha \tag{3}$$

where higher order terms in k and $\alpha$ are neglected.

The scattered electric field modulation $\Delta E_s$ is directly proportional to $\Delta\alpha$; one therefore expects to see a strong decrease in $\Delta E_s$ as the tip-feature dipole-dipole coupling decreases with increasing r. Figure 5 gives a plot of the optical signal as a function of spacing, compared with the theoretical curve obtained from equation (1). The sample was a chromium film, acting as a mirror, and the dipole-dipole interaction consisted of the tip interacting with its image. We observe a strong decrease in the optical signal over a distance corresponding to the tip diameter, 370 Å in this case. Furthermore, equation (1) shows that $\Delta\alpha$ is proportional to the product of the complex polarizability of the tip $\alpha_t$ and that of the feature $\alpha_f$. Consequently, the phase of the scattered field component $\Delta E_s$ can change drastically depending on the complex polarizability of the tip-end as observed in Fig.4.

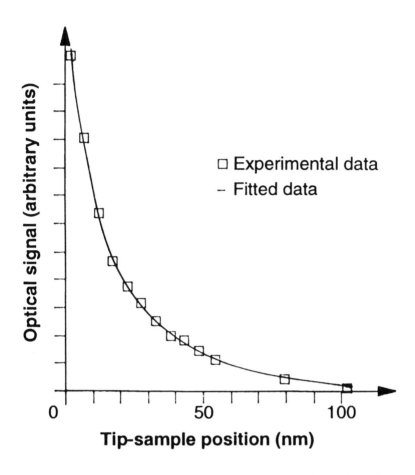

Fig.5    Measured optical dipole-coupling signal (after subtraction
        of instrument response) versus tip-sample spacing,
        compared with theory.

In general, the scattered field will have a component $E_{s\phi}$ which is orthogonal in phase to $E'_r$ and provides a phase shift, and a component $E_{se}$ out of phase with $E'_r$ which provides an extinction in the interferometer. Combining equations (1) and (2) with the far-field expression for $E'_r$, $E'_r = \dfrac{E_i \lambda n}{i \, 5 \, \pi \, d \, NA^2}$ (9), and noting that the polarizability is related to the susceptibility $\chi$ by $\alpha = \dfrac{\chi}{4\pi} \left( \dfrac{4}{3} \pi a^3 \right)$, expressions for the phase shift and extinction modulation in the interferometer are:

$$\frac{\Delta E_{se}}{E'_r} = \frac{5}{9} \, (ka)^3 \, \frac{NA^2}{n^2} \, Im[\chi_t \chi_f] \tag{4}$$

$$\frac{\Delta E_{s\phi}}{E'_r} = \frac{5}{9} \, (ka)^3 \, \frac{NA^2}{n^2} \, Re[\chi_t \chi_f] \tag{5}$$

## 5. Conclusion

Several conclusions can be inferred from these equations. First, both the *real* and *imaginary* parts of the susceptibility of a feature could be determined - in principle down to the atomic (9) scale - with two simultaneous measurements of in-phase and quadrature components; tip susceptibility being measured independently using a known reference surface as the sample. Experimentally, absolute values of interferometric signals, and their variation with tip susceptibility (a factor 4x increase for Al versus Si tip) agree with equations (4) and (5).

Another key advantage of the interferometric measurement in SIAM is the detection of the scattered electric fields that varies as $(ka)^3$ rather than intensities that vary as $(ka)^6$ as in typical near-field optical systems. The interferometric signal is significantly higher than background noise sources (such as stray light, dark current or thermal noise), which are inevitably present in experimental set-ups. By contrast, this background noise severely limits the sensitivity of typical NSOM at high resolution (smaller than 50 nm). The understanding of the principles involved in SIAM opens up many capabilities for imaging and spectroscopy at sub-nanometer scales and particularly extends its application to biology.

# 6. References

1.  Ash, E.A and Nicholls, G.(1972), *Nature.*, **237**, 510

2.  Pohl, D.W., Denk, W., and Lanz, M.(1984) *Appl. Phys. Lett.*, **44**,651

3.  Lewis,A., Issacson, M., Haratounian, A., and Murray, M.(1984) *Ultramicroscopy*, **13**, 227

4.  Fee, M., Chu, S., and Haensch, T.W.(1989) *Optics Com.*, **69**(3,4),219

5.  Betzig, E and Trautman, J.(1992) *Science*, **257**, 189

6.  Toledo-Crow, R., Tang P.C., Chen, Y., and Vaez-Iravani, M.(1992) *Appl. Phys.Lett.*, **60**(24), 2957

7.  Betzig, E and Chichester, R.J.(1993) *Science*, **262**, 1422

8.  Wickramasinghe, H.K., and Williams, C.C., Apertureless Near Field Optical Microscope, US Patent 4,947,034 (April 28, 1989); IBM Disclosure #Y0887-0949, Dec 1987

9.  Zenhausern, F., O'Boyle, M.P., and Wickramasinghe, H.K.(1994) *App.Phys.Lett.*, **65**, 1623

10. FITC latex spheres were provided by Polysciences Inc., lot#17151

11. Martin, Y., Williams, C.C., and Wickramasinghe, H.K.(1987) *J.Appl.Phys.*, **61**(10), 4723

12. Jackson, J.D.(1957) *Classical Electrodynamics* , (Wiley, New York)

13. van de Hulst, H.C.(1957) *Light Scattering by Small Particles* , (Wiley, New York)

14. Zenhausern, F., Martin, Y., and Wickramasinghe, H.K.(1995) *Science*, **269**, 1083

15. F.Z. was supported by a SNF grant # 5002-36851

# PRIMARY IMAGING MODES IN NEAR-FIELD MICROSCOPY

M. VAEZ-IRAVANI
*Tencor Instruments,*
*2444 Charleston Road,*
*Mountain View,*
*CA 94043,*
*USA.*

## 1. Introduction

Near field scanning optical microscopy in its various modalities has now reached a stage where it can be considered a powerful, main-stream, analytical tool {1-16}. The basic promise of the technique, viz. providing optical contrast at a resolution below the classical diffraction limit, has been realized in a variety of settings. The contrast mechanisms in the near-field microscopy can, broadly speaking, be divided into two major categories. In the first, one finds those effects that are inherent in the wave nature of light, such as amplitude, phase, and polarization. In this class, the interaction of the light with the sample does not involve any transformation of the nature of the radiation, and the imaging process is one of registering, with a given accuracy, the modulation imparted to the light. In the other category there exist a variety of mechanisms, such as fluorescence, photoluminescence and Raman imaging and spectroscopy, effects primarily based on the particle nature of light. There has been impressive success in capturing such effects, and no doubt such work will continue to provide fruitful results.

The main purpose of this paper is to describe the primary contrast mechanisms in near-field microscopy (NSOM), namely amplitude, phase and polarization. Here, we address the use of the aperture NSOM, and describe some of the basic challenges facing near field microscopy. We note, however, the impressive results obtained with the other types of near-field systems, including, in particular, the apertureless NSOM {17}, and integrated detector approaches {18}.

## 2. Amplitude Imaging in Near Field Microscopy

A major impediment to the rapid progress of the field was, for a long time, the lack of a reliable, and universally acceptable, distance regulation for the tip/sample separation. The force regulation of this gap {19-22} appears to have successfully resolved this issue in most applications. The basic premise here is that the local topography of the sample alters the free-oscillation amplitude of the tip which, when properly utilized in a feed-

*M. Nieto-Vesperinas and N. García (eds.), Optics at the Nanometer Scale* 143–150.

back circuit, amounts to tracking of the topography. Indeed, a majority of NSOM's reported to date utilize this method. Furthermore, the detection of the tip vibration has mostly been carried out optically. More recently, however, capacitive {23} and piezo-electric {24} techniques have also been reported. These methods have the potential benefit that the force regulation detection does not interfere with light sensitive samples.

Notwithstanding the above, it is important to note that a topographic interpretation of the signals obtained inherently negates other possible mechanisms. It can be shown, for example, that the surface affinity for contaminations, especially, its hydrophobicity/hydrophilicity can have a profound effect on the relative value of a given topographic step on the signal {25}. The overall conclusion seems to be that the assignment of a topographic interpretation to a given signal is robust, as long as one is reasonably assured of the relative significance of these effects. The technique is non-contact and non-invasive, provided that the amplitude of the tip sideways oscillations is kept small (a few nanometers at most), and care is exercised in the initial approach to the sample.

Regardless of the method of distance regulation, the tip can be used either to illuminate the sample, or to detect the light emitted by the sample, so as to form a near-field optical image. Unfortunately, this immensely significant ability to have a simultaneous image of the topography and optical characteristics also fuels one of the most controversial aspects of force-regulated NSOM, namely the coupling between the optical and force modalities. This coupling is often manifested in the appearance of high resolution (sometimes extremely high resolution, well beyond that expected from the aperture size considerations) features in the optical image. The origins of this anomaly can be traced to one, or a combination, of the following effects: i) protrusions in the tip area; ii) the difference between the optical and the force approach curves; iii) the appropriate settings for the feedback parameters. It is, therefore, crucial that one obtains and compares the force and optical images (even if the former is later discarded) to ensure the validity of any claim about resolution.

Near-field optical imaging can be carried out in transmission and reflection, with the latter being the more problematic of the two modalities. One of the difficulties of the reflection approach is the sample induced shadowing in the oblique geometry {26}, and the self-shadowing effect of the tip in the normal direction. The collection efficiency in this mode is also, in general, less than that in the transmission mode. In addition, both narrow band (eg laser) and broad-band light sources can be used for illumination proposes. In the broadband case, one uses a multi-mode fiber link to bring the radiation to the tip region. A comparison of the results obtained using these two illumination sources provides an excellent demonstration of the effect of temporal coherence in near field microscopy.

## 3. Phase Contrast Microscopy in the Near Field

Perhaps the most direct method of the extraction of the phase of light is to resort to interferometry. In the near field such a system was first demonstrated in a pseudo-heterodyne approach, using a modified Mach-Zehnder configuration {27}. A basic problem with all separate-beam interferometers, however, is the fact that the two arms experience different spurious phase fluctuations due to microphonics. When fiber links are used, as in the case of near-field microscopy, such effects are dramatically exacerbated. It is, therefore, necessary to introduce a correction mechanism to account for the instability in the interference signal. Indeed, this task was performed in the above system using the technique of second harmonic cancellation in a feedback system. The apparent complication of the introduction of a second feedback circuitry in the system can, in fact, be used to advantage so as to establish different regimes of operation: pure amplitude, phase contrast, and open-loop. A potentially important application of interference near-field microscopy is in the study of the refractive index variations in samples, where the extinction coefficient is not sufficient, and contrast enhancing agents are not available, to allow efficient amplitude-only microscopy. It is expected that the benefits of this approach in such cases will parallel those attained in far field phase contrast microscopy, such as the Zernike method.

Yet, an unorthodox use of this near-field interference contrast technique is in the high sensitivity detection of Faraday rotation, such as that encountered in magneto-optic systems. Consider, for example, a sample that imparts a certain rotation of polarization to the probing light. If we arrange for the emerging light to be collinear with respect to a reference beam, the two beams will interfere provided that they are not orthogonally polarized. Thus, if in the absence of a sample-induced rotation of plane of polarization the two beams are orthogonally polarized, the presence of any interference signal is necessarily an indication of polarization re-orientation of at least one beam. In practice, the sample beam is phase modulated, and the detection is performed at the modulation frequency. Due to the pseudo-heterodyne nature of the detection, this method provides shot-noise limited sensitivity signals which are linearly dependent on the polarization rotation {28}.

## 4. Polarizing Near-Field Microscopy

The transduction process in near-field microscopy involves the passage of light through a small, non-guiding, aperture. It is, perhaps, quite remarkable that the polarization of the light exiting a symmetric aperture is essentially preserved. Theoretical considerations {29} indicate that linearly polarized launched light gives rise, at the aperture, to certain amounts of the two other orthogonal polarizations. The relative value of these other polarizations rises as the aperture shrinks in size. The preservation of the polarization by the near-field probe means that the technique is inherently able to perform sub-diffraction limit polarization imaging.

It follows from the above that the experimental implementation of near-field polarizing microscopy can parallel that in far-field systems, viz. viewing the object between a pair of crossed polarizers. This, in fact, was the first mode of near-field polarizing microscopy {30-31}. On the other hand, as a high resolution analytical tool for birefringence assessment, the value of near field polarizing microscopy is substantially enhanced when one is able to draw conclusions about the nature of the birefringence in the samples under examination. Such analyses require a methodology for the preservation of the sign of the sample birefringence, which is afforded by resorting to a pure linear polarization imaging system {32}. In this case, at any given moment in time, two simultaneous AC signals are generated, whose ratio gives a highly sensitive, pure birefringence signal.

A basic requirement for any polarizing microscope is stability. In the near field microscope this problem is more acute, as the fiber link renders the system more susceptible to temperature induced fluctuations. If, in addition, one is imaging a low birefringence sample, it becomes necessary to reduce the scan speed, and the bandwidth, in an effort to increase the S/N. The prolonged scan time can place serious limitations on the performance of the system due to the environmental fluctuations. To alleviate such problems, the fiber used in the system is preferably a polarization preserving one, and the light can be launched into one of the axes of the fiber. The phase modulation system to effect the linearization can then follow the sample {33}. Finally, one can use a symmetric aperture at the tip of a polarization preserving fiber as a passive probe to investigate the focal distribution of lenses. Indeed, such an arrangement has been used to determine the distribution of y-polarized light at the focus of an x-polarized illuminated, high numerical aperture, lens {28}.

## 5. Representative Results

Examples of the results obtained with various primary imaging modalities of the NSOM abound in the literature. Here, we present a few illustrative examples, again only to show the primary contrast modes. Fig. 1 (a), (b) and (c) show the application of direct imaging NSOM in probing line structures on the small scale: (a) is the transmission image of irregular lines in chrome (3 $\mu$m field of view), and (b) and (c) are reflection coherent and incoherent images of poly-silicon lines on Si substrate (7 $\mu$m on the side) {26}. Fig. 1(d) is a linear birefringence image of 200 nm thick Kevlar-29 sample, showing the characteristic butterfly pattern, together with a number of parallel lines across the sample. These were confirmed to be due to the microtoming process, by comparison with the simultaneously obtained force image (not shown here) {33}. The field of view is 9 $\mu$m. Fig. 1(e) shows the rectified pure polarization image of a number of nematic liquid crystal (LC) globules in a polymer matrix. The corresponding topography image is presented in 1(f), showing that, in fact, the majority of the LC structures are located below the surface (field of view is 9 $\mu$m). Fig. 1(g) is a near-field, pseudo-heterodyne interferometrically detected collection of conventionally written bits on a magneto-optic substrate. Finally, Fig. 1(h) shows the y-polarized focal distribution, detected using a near-field probe, due to an x-polarized input laser beam into a high numerical aperture lens {28}.

## 6. Discussion and conclusions

The primary motive for resorting to NSOM is, of course, the attainment of high resolution. Clearly the NSOM has demonstrated its potential in providing enhanced resolution. The confined illumination afforded by the small aperture also has the advantage that at any given time only the relevant parts of the sample are irradiated. This

is an important asset in a number of experiments where the illumination can have a permanent effect on the sample, such as photo-bleaching of fluorescent agents. On the other hand, one has to consider these advantages against the drawbacks of the system. Chief amongst these are: i) the attenuation of light upon passage through the aperture (of the order of $10^6$:1); ii) the heating effect due to the loss of light at the tip {34-35}; iii) possible quenching of fluorescence due to the proximity of the metallized tip and the sample {12}; iv) the restriction of the probed region to the surface or near-surface parts of the sample.

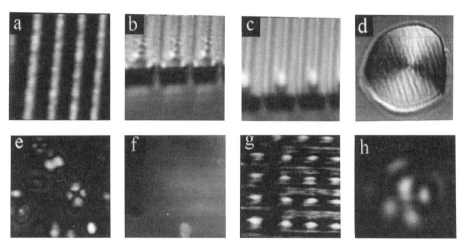

Fig. 1. Sample of different NSOM applications (primary contrast mechanisms only; see text for details)

It appears that in the next phase of the development of near-field microscopy attention must be paid to those aspects of the technique which provide a unique imaging capability. To illustrate, let us note that many of the interesting results that have been obtained with the NSOM have been taken with a resolution in the range of 70-100 nm. This is substantially superior to that available in the current far-field systems. However, recent work in the latter area both in the single {36}, as well as the two-photon processes {37}, have also brought the "conventional" resolution limit to within the striking distance of this range. Under these circumstances, the use of near field microscopy must be evaluated against the various disadvantages of the technique. It appears logical that attention should now be focused on the further exploitation of the simultaneity of contrast mechanisms afforded by the technique, or on the substantial enhancement of the resolution to offset the other complications associated with the technique. Recent work on the apertureless NSOM holds a great promise in this regard {17}. At the same time, the availability of the topographic information as a separate entity in NSOM imaging is extremely powerful, and is likely to remain a major reason for the continued interest in NSOM. In addition, recent theoretical work {38} have indicated the possibility of extending the application of near-field microscopy to true surface reconstruction, and hold promise for major new uses in the future.

## Acknowledgements

I am grateful to R. Toledo-Crow, Y. Chen, J.K. Rogers, and D.I. Kavaldjiev, my former students at the Center for Imaging Science, Rochester Institute of Technology, with whom this research was carried out. I am also grateful to H. Ade, and R. Spontak, of North Carolina State University for collaboration on the Kevlar imaging work. This work was supported by grants from the Whitaker Foundation, and the National Science Foundation.

## References

1. Pohl, D.W., Denk, W., and Lanz, M. (1984) Image recording with resolution $\lambda/20$, *Appl. Phys. Lett.* **44**, 652-653.
2. Lewis, A., Isaacson, M., Haootunian, A., and Muray, A. (1984) Development of a 500 A spatial resolution light microscope, *Ultramicroscopy* **13**, 227-232
3. de Fornel, F., Goudonnet, J.P., Salomon, L., and Lesniewska, E. (1989) An evanescent field optical microscope, *SPIE* **1139**, 77-79
4. Cline, J.A., Barzshatsky, H., and Isaacson, M. (1991) Scanned tip reflection mode near-field scanning optical microscopy, *Ultramicroscopy* **38**, 299-304.
5. Vigoureux, J.M., Girard, C., and Courjon, D. (1989) General principle of scanning tunneling optical microscopy, *Opt. Lett.* **14**, 1039-1041.
6. Fischer, U. Ch., Durig, U.T., and Pohl, D.W. (1988), *Appl. Phys. Lett.* **52**, 249-251.
7. Paesler, M.A., Moyer, P.J., Jahncke, C.J., Johnson, C.E., Reddick, R.C., Warmack, R.J., and Ferrel, T.L. (1990) Analytical photon scanning tunneling microscopy, *Phys. Rev. B* **42**, 6750-6753.
8. Betzig, E., Trautman, J.K., Harris, T.D., Weiner, J.S., and Kostelak, R.L. (1991) Breaking the diffraction barrier: optical microscopy on a nanometric scale, *Science* **251**, 1468-1470.
9. van Hulst, N.F., Sergerink, F.B., Achten, F., and Bolger, B. (1992) Evanescent field optical microscopy: effects of polarization, tip shape and radiative waves, *Ultramicroscopy* **42**, 416- 421.
10. Fischer, U. Ch., and Zapletal, M. (1992) The concept of a coaxial tip as a probe for scanning near-field optical microscopy and steps towards a realization, *Ultramicroscopy* **42**, 393-398.

11. Hess, H.F., Betzig, E., Harris, T.D., Pfeiffer, L.N., and West, K.W. (1994) Near-field spectroscopy of the quantum constituents off a luminescent system, *Science* **264**, 1740-1745.

12. Xie, X.S. and Dunn, R.C. (1994) Probing single molecule dynamics, *Science* **265**, 361-364.

13. Ambrose, W.P., Goodwin, P.M., Martin, J.C., and Keller, R.A. (1994) Alteration of single molecule fluorescence lifetimes in near-field optical microscopy, *Science* **265**, 364-367.

14. Trautman, J.K., Macklin, J.J., Brus, L.E., and Betzig, E. (1994) Near-field spectroscopy of single molecules at room temperature, *Nature* **369**, 40-42.

15. Rogers, J.K., Seiferth, F., and Vaez-Iravani, M. (1995) Near-field probe microscopy of porous silicon: observation of spectral shifts in photoluminescence of small particles, *Appl. Phys. Lett.* **66**, 3260-3262.

16. Rogers, J.K., Toledo-Crow, R., Vaez-Iravani, M. DiFrancisico, G., Zhao, T., and Hailstone, R. (1995) Correlative near-field direct/fluorescence imaging and spectroscopy of a sensitizing dye on single microcrystals of silver halide, *Jour. Imag. Sci. Tech.* **39**, 205.

17. Zenhausern, F., O'Boyle, M.P., and Wickramasinghe, H.K., Apertureless near field optical microscope, *Appl. Phys. Lett.* **65**, 1623-1625.

18. Davis, R.C., Williams, C.C., and Neuzil, P. (1995) Micromachined submicrometer photodiode for scanning probe microscopy, *Appl. Phys. Lett.* **66**, 2309-2311.

19. Yang, P.C., Chen, Y., and Vaez-Iravani, M. (1992) Attractive mode force microscopy with optical detection in an orthogonal cantilever/sample configuration, *Jour. Appl. Phys.* **71**, 2499.

20. Betzig, E., Finn, P.L., and Weiner, J.S. (1992) Combined shear force and near-field scanning optical microscopy, *Appl. Phys. Lett.* **60**, 2484-2486.

21. Toledo-Crow, R., Yang, P.C., Chen, Y., and Vaez-Iravani, M. (1992) Near-field differential scanning optical microscope with atomic force regulation, *Appl. Phys. Lett.* **60**, 2957-2959.

22. van Hulst, N.F., Moers, M.H.P., Noordman, O.F.J., Faulkner, T., Serink, F.B., van der Werf, B., de Grooth, G., and Bolger, B. (1992) Operation of a scanning near field optical microscope in reflection in combination with a scanning force microscope, *SPIE* **1639**, 36-43.

23. Leong, Y. and Williams, C.C.(1995) Shear force microscopy with capacitance detection for near-field scanning optical microscopy, *Appl. Phys. Lett.* **66**, 1432-1434.

24. Karrai, K. and Grober, R.D. (1995) Piezoelectric tip-sample distance control for near field optical microscopes, *Appl. Phys. Lett.* **66**, 1842-1844.

25. van Hulst, N.F., *Ultramicroscopy* (in press)

26. Toledo-Crow, R., Smith, B.W., Rogers, J.K., Vaez-Iravani, M. (1994) Near-field optical microscopy characterization of IC metrology, *SPIE* **2196**, 62-73.

27. Vaez-Iravani, M. and Toledo-Crow, R. (1993), Phase contrast and amplitude pseudo-heterodyne interference near-field scanning optical microscopy, *Appl. Phys. Lett.* **62**, 1044-1046.

28. Toledo-Crow, R., Rogers, J.K., Seiferth, F., and Vaez-Iravani, M. (1995) Contrast mechanisms and imaging modes in near field optical microscopy, *Ultramicroscopy* **57**, 293-297.

29. Bouwkamp, C.J. (1950) On Bethe's theory of diffraction by small holes, *Philips Res. Res.* **5**, 321-332.

30. Betzig, E., Trautman, J.K., and Weiner, J.S. (1992) Polarization contrast in near-field scanning optical microscopy, *Appl. Opt.* **31**, 4563-4568 (1992)

31. Betzig, E., Trautman, J.K. Wolfe, R., Gyorgy, E.M., Finn, P.L., Kryder, M.H., and Chang, C.H. (1992), Near-field magneto-optic and high density data storage, *Appl. Phys. Lett.* **61**, 142-144.

32. Vaez-Iravani, M. and Toledo-Crow, R. (1993), Pure linear polarization imaging in near-field scanning optical microscopy, *Appl. Phys. Lett.* **63**, 138-140.

33. Ade, H., Toledo-Crow, R., Vaez-Iravani, M., and Spontak, R.J., Observation of polymer birefringnce in near-field optical microscopy, *Langmuir* (in press)

34. Kavaldjiev, D.I., Toledo-Crow, R., and Vaez-Iravani, M. (1995) On the heating of the fiber tip in a near-field scanning optical microscope, *Appl. Phys. Lett.* **67**, 2771-2773.

35. LaRosa, A.H., Yakobson, B.I., and Hallen, H.D. (1995) Origins and effects of thermal processes on near-field optical probes, *Appl. Phys. Lett.* **67**, 2597-2599.

36. Vaez-Iravani, M. and Kavaldjiev, D.I. (1995) Resolution beyond the diffraction limit using frequency-domain field confinement in scanning microscopy, *Ultramicroscopy* (in press)

37. Hell, S. (1994) Improvement of lateral resolution in far-field fluorescence light microscopy by using two-photon excitation with offset beams, *Opt. Com.* **106**, 19-24.

38. Garcia, N. and M. Nieto-Vesperinas (1995) Direct solution to the inverse scattering problem for surfaces from near-field intensities without phase retrieval, *Opt. Lett.* **20**, 949-951.

# LOCAL EXCITATION OF SURFACE PLASMONS BY "TNOM"

B. HECHT, D. W. POHL
*IBM Research Division, Zurich Research Laboratory,*
*8803 Rüschlikon, Switzerland*

AND

L. NOVOTNY
*Institut f. Feldtheorie und Höchstfrequenztechnik, ETH Zürich,*
*CH-8092 Zürich, Switzerland*

**Abstract.** A point light source, such as the aperture scanning near-field optical microscope (SNOM) probe in emission mode, can excite surface plasmon polaritons (SP) in metallic films by means of its near field. We demonstrated this using a *tunnel near-field optical microscope (TNOM)* with thin silver films as samples. Light is transmitted through the film predominantly into the directions defined by the dispersion relation of the SP, forming a sharply confined conical sheet of radiation in the far field. The angle of emission is larger than the critical angle $\theta_c$ for total reflection, hence not accessible to standard SNOM. The weak emission into allowed directions mainly stems from SP scattering at film imperfections.

With TNOM, it is possible to record "forbidden" ($\theta \geq \theta_c$) and "allowed" ($\theta \leq \theta_c$) light separately but simultaneously. The resulting scan images represent the intensities of SP *excitation* and SP *scattering* at imperfections of the silver film as a function of probe position. Their most conspicuous features are concentric circles and confocal hyperbolae, respectively, suggestive of *interactions between plasmons* originating from the tip and from nearby scattering centers. Our findings indicate that plasmon scattering/reflection can be studied quantitatively and on a local scale.

## 1. Introduction

The properties of surface plasmon polaritons (surface plasmons for short, or SP) in thin metallic films have been studied extensively [1]. Their prop-

*M. Nieto-Vesperinas and N. García (eds.), Optics at the Nanometer Scale 151–161.*

erties with regard to dielectric constant $\epsilon$, thickness, substrate, etc. are well understood. Less well understood, although of relevance for technical applications, is the influence of inhomogeneities, such as the reflection at the border of a film, the scattering at an imperfection, or the coupling to an adsorbate at the film surface.

With the advent of scanning near-field microscopy, it became possible to study plasmons on a very local scale. The near-field optical (NFO) probe can act as a local SP resonator [2], scattering center [3-5], detector [6-10], or — when used in emission mode — as *point source* of "plasmon" waves.

The near field at the exit plane of an illuminated subwavelength-diameter aperture [11, 12] is confined to such a small spot size that its spatial spectrum extends over a large range of $k$-vectors, including those that match the $k$-vector of a freely propagating SP in a metal film, as sketched in Fig. 1.

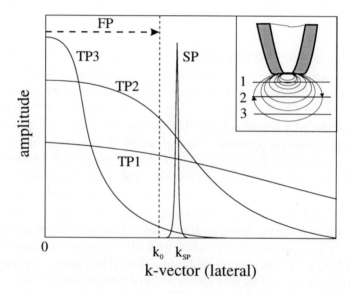

*Figure 1.* Amplitudes vs. component of the $k$-vector in the film plane for various types of excitation, schematic: FP: freely propagating photon, increasing angle of incidence; TP: photons tunneling out of aperture, producing the evanescent wave on the exit side, 1 to 3 correspond to increasing distance; SP: thin-film surface plasmon for given frequency of excitation. Inset: End piece of the NFO probe; the circular lines give a suggestion of the electric field next to the aperture; cuts at levels 1–2–3 produce spots of increasing diameter, corresponding to curves TP1–TP3 in the lateral $k$-area.

Emission-mode SNOM thus provides an alternative to the commonly used techniques of Kretschmann and Raether [13] and Otto [14]. Unlike in these two methods, emission-mode SNOM excites circular SP waves with well-defined origin and confined to an area around the NFO probe the radius of which is given by the SP damping length. This allows the observation of

SP effects within a selected, microscopic-size regime, which is of particular interest for the study of interactions with individual perturbations of the metal film.

Here we report on the experimental observation of SP *excitation* by a SNOM aperture probe, on local variations of the coupling efficiency to a thin silver film, and on the influence of the gap width. Scan images obtained by the *tunnel near-field optical microscope TNOM-II* [15, 16], in particular, provide somewhat surprising new insights into the mechanisms of plasmon *interference* and *scattering*.

## 2. Experimental

The TNOM is a combination of an aperture-SNOM [11, 17, 18] and an inverted scanning tunneling optical microscope (STOM [19] or photon scanning tunneling microscope "PSTM" [20]). It provides simultaneous records of the light transmitted from the NFO probe into the so-called *forbidden* and *allowed* directions. The two regimes are separated by the critical angle for total internal reflection $\theta_c = \arcsin(1/n) = 41.12°$, where $n = 1.51$ is the refractive index of the film glass substrate.

The two setups used in our experiment are shown in Fig. 2. Common to both are NFO probes (NFO-p), pulled and coated glass fiber tips with aperture, and the light source, an argon laser operated at $\lambda_0 = 514.5$ nm.

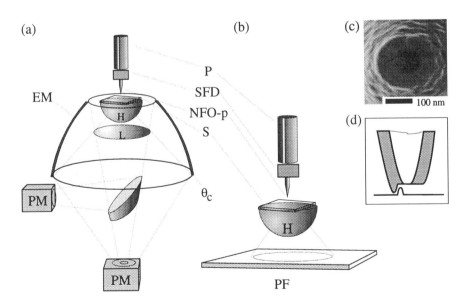

*Figure 2.*   TNOM: (a) original setup (from Ref. [16]), (b) setup for the investigation of the angular distribution, (c) SEM image of the apex of an NFO probe, (d) schematic sideview of an NFO probe at operating distance from the sample surface

The state of polarization can be adjusted by a "fiber-loop" polarization controller [21]. An end-on SEM image of an aperture NFO probe used in our experiments is given in Fig. 2c. The aperture shows up as a dark spot in the center, with a diameter of $\approx 80$ nm.

Mounted on a 1"-long, 1/4"-wide piezotube (P), the NFO probe can be scanned over an area of about $10 \times 10\,\mu^2$. The gap width is regulated by "shear" force detection (SFD) with a miniaturized fiber-optic interferometer [16], similar to the one developed by Rugar et al. [22]. This method allows a very small vibration amplitude, typically 1 to 2 nm ptp. The aluminum coating of the NFO probe is not completely flat but consists of slightly protruding grains (crystallites), 20 to 100 nm in diameter. Therefore, it has to be assumed that the mechanical point of contact between tip apex and object surface is at the rim of the aperture, i.e., displaced from the optical center of the probe, as sketched in Fig. 2d.

The samples (S) are silver films of nominal 50 nm thickness, deposited by evaporation at $5 \times 10^{-6}$ mbar on thin cover glasses. The thickness chosen is optimal for plasmon experiments [1]. This can be verified by monitoring the reflectivity in the Kretschmann configuration: The reflected radiation into the direction of the resonance angle $\theta_{SP}$, known from literature data [1], must be completely quenched in this case, a condition that was checked for each of sample prior to installation in the TNOM. The sample is mounted in optical contact with a BK7 glass hemisphere (H), which transmits all the light emerging from the bottom of the sample up to $\theta = 90°$.

Figure 2a shows the TNOM version used for scan imaging and quantitative gap-width control [16]. Forbidden and allowed radiation fall onto an elliptical mirror (EM) and an objective (L), respectively, which focus them on separate photomultipliers (PM). In the second version, Fig. 2b, the collective optics is replaced by a Polaroid cassette (PF) in order to study the angular distribution of the transmitted light. Allowed and forbidden directions correspond to the inside and outside areas of a circle with radius $r_c = a \cdot \sin\theta_c = a/n$, where $a$ is the distance of the film from the NFO probe.

## 3. Characteristics of Plasmon/Light Coupling

### 3.1. ANGULAR DISTRIBUTION

To clearly distinguish between the plasmon effects of interest and instrumental properties, the distribution of the transmitted radiation was photographed first with a bare cover glass as sample.

In Fig. 3a, the NFO probe is still several microns away from surface of the hemisphere. The transmitted radiation is sharply confined to the cone of allowed directions, $\theta \leq \theta_c$, as expected from the laws of classical

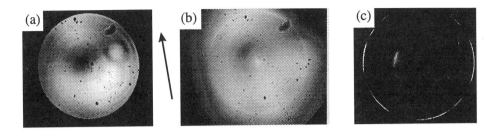

*Figure 3.* Angular distribution of transmitted radiation: (a,b) without object, NFO probe in retracted and operating positions, respectively; (c) silver film, operating position.

optics. The structure visible inside the circle is oriented along the direction of polarization indicated by the arrow. The asymmetry of the two lobes is caused by imperfect alignment of the probe with regard to the center of the hemisphere; the other details in the light distribution, probably caused by small instrumental imperfections, are not of interest here.

In Fig. 3b, the NFO probe is at operating distance, defined by the onset of frictional interaction between NFO probe and sample surface. The illuminated area has become larger, and the borders are diffuse; the $r_c$ borderline circle can no longer be recognized. The extra light represents the forbidden radiation. Its total flux approximately equals that into the allowed directions. We found that the flux ratio (forbidden/allowed) provides a rough measure for the size of the aperture: It increases markedly with shrinking diameter.

With the silver film as sample and the NFO probe in retracted position, the transmission was too low for a photographic record. During approach, however, a well-defined ring structure began to shine, went through an intensity maximum (cf. Sect. 3.2) and finally ended up with a 20–30% lower brightness at operating distance, shown in Fig. 3c. The radius $r \geq r_c$ corresponds to $\theta = 44.3°$, which can readily be identified as the resonance angle $\theta_{SP}$ for the given frequency of excitation [1]. The corresponding plasmon wavelength is $\lambda_{SP} = 488$ nm. Some of the gaps in the ring structure correlate with the direction of polarization; the others might reflect imperfections of the NFO probe and/or the silver film. The faint, diffuse radiation background in the allowed regime may in part result from direct transmission and also from plasmon scattering at the various little imperfections interspersed in the silver film (see Sect. 4.1).

## 3.2. INFLUENCE OF GAP WIDTH

The variation of the signal with the gap width is an important parameter for any scanning probe microscope. Figure 4 shows the photomultiplier signals

(a)                           (b)

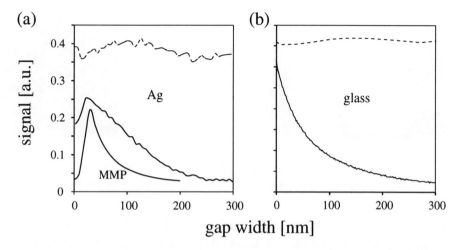

gap width [nm]

*Figure 4.* (a) "Approach curves". Solid/dashed lines: forbidden/allowed light; Ag: silver film; glass: bare cover slide; MMP: numerical simulation.

for the final 300 nm of approach for the silver film and for a bare glass slide. The silver film thickness was less than optimum in this case, resulting in sizeable direct transmission of allowed light. The features which we found to be characteristic for the SP/photon interaction still are clearly visible.

At large gap width, both signals from the silver film more or less exhibit the well-known standard distance dependence [16]: The allowed signal is slightly modulated with a period of $\lambda_0/2$; the forbidden signal becomes exponentially smaller with a decay constant of about 120 nm. At a gap width of about 30 nm, the exponential curve comes to a sudden end, and the intensity decreases markedly until a minimum is reached at $\approx 10$ nm, which approximately equals the operating distance. In the allowed light, a small minimum at approx. 30 nm gap width is registered, but no significant deviation from the general undulatory character of the approach curve is found.

The decrease in coupling efficiency at small gap width does not exist for the bare glass surface (see in Fig. 4b), hence has to be considered typical for plasmon excitation. This assumption is supported by a numerical calculation of the "approach curve" (trace "MMP" in Fig. 4a) with a realistically modeled aperture NFO probe interacting with a 50-nm-thick silver film [23, 24], cf. Fig. 6a.

The unexpected finding of an optimum gap width for SP excitation may be the result of several effects, in particular:

1. The light intensity at the silver surface and within the film increases drastically with decreasing gap width owing to the presence of photons that tunnel out of the aperture but do not propagate into the far

field — *this is the description of the "evanescent wave" in the particle picture*. At the same time, the lateral extension shrinks, corresponding to an increase of the tunnel photon (TP) density with the lateral $k$-vector equaling $k_{SP} = 2\pi/\lambda_{SP}$. The situation is sketched in Fig. 1 for three distances 1–2–3. In real space (inset of Fig. 1), it corresponds to the optimization of the spot diameter with respect to $\lambda_{SP}$, which is intuitively expected for $\lambda_{SP}/2$.

2. In the immediate proximity of the NFO probe, the SP dispersion will be influenced by the presence of the aluminum cladding. If the extension of this zone were large compared to $\lambda_{SP}$, the speed of SP propagation would differ from that of the unperturbed film, resulting in some degree of reflection at the borderline. In our experiment, however, the size of the interaction zone is not much larger than $\lambda_{SP}/2$, hence represents a kind of *SP quantum dot* with distinct, though very broad, resonance frequencies. The fixed excitation frequency will in general not match the levels of the SP quantum dot, hence couple less and less efficiently with shrinking gap width.

3. A third possible mechanism is that the geometry of the probe might favor the emission of light generated by the SP into the air space to the sides of NFO probe, but so far we have found no evidence of such a behavior.

At the current state investigations, we cannot yet decide which is the dominant mechanism for the decay of coupling efficiency.

## 4. TNOM images

### 4.1. DESCRIPTION

It is clear from Fig. 3c that the forbidden image will essentially represent the intensity of the plasmon resonance radiation and its variation with probe position: The allowed image will reflect the corresponding variation of the background radiation. The SFD mechanism in addition provides a topographic picture of the sample surface for each of the optical images.

Figure 5 depicts a set of such pictures that show the peculiar plasmon effects to be reported here particularly well. Scan range and speed were $6\mu \times 6\mu$ and 1 line/sec, respectively. The $256 \times 256$ pixel image was obtained in about four minutes. To start with the topography, Fig. 5a, the film appears to be flat within about 1 nm and unstructured except for a number of tiny protrusions, $\leq 150$ nm in diameter and $\leq 40$ nm in height, which probably are growth hillocks (their exact nature is irrelevant for this discussion).

In the optical images, Fig. 5b and c, the most conspicuous features are the periodic intensity variations resembling interference patterns. It will be shown, however, that they are not interference patterns in the common

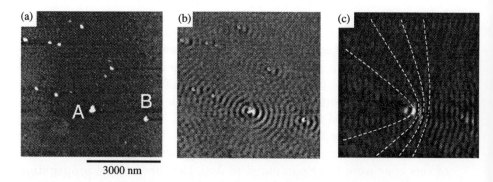

*Figure 5.* Scan images of a 60 ± 20-nm-thick silver film. (a) Topography, (b) forbidden and (c) allowed TNOM images. In (c) the family of confocal hyperbolae with foci at A,B has been indicated. The branches near A fit the dark fringes over sizeable distance.

sense. The protrusions are clearly visible in both optical images. They appear to be slightly displaced with regard to the centers of the undulation structures. In the forbidden image, the undulations surround the protrusions in the shape of circular rings. In the allowed image, some of the fringes can be fitted quite well to the confocal hyperbolae marked by dashed lines. Their centers of eccentricity are at A,B, and the mutual distance is $\lambda_{SP}/2$ on the A–B axis.

## 4.2. DISCUSSION

When deposited on a glass substrate, thin-film SPs lose energy by emission of light at a large, well-defined angle $\theta_{SP}$, the value of which can be determined from the SP and photon dispersion relations [1]. The SP wave can still propagate over considerable distances in the silver film. Typical decay lengths are of the order $20\,\mu$ [1]. The emitting area hence is by no means as confined as the aperture of the NFO probe and not even as the period of undulation seen in Figs. 5b and 5c. Thus, the forbidden signal is a measure for the amplitude of the overall SP excitation. Variations during scanning have to be interpreted as changes in the photon/SP coupling efficiency.

The loci of equal intensity in the forbidden image, Fig. 5b, are concentric circles with their centers at or near (see below) the protrusions. The azimuthal variation of the amplitude corresponds to the dominant direction of polarization of the highly polarized radiation from the NFO probe. The radial period of undulation is $240\pm 5$ nm, exactly one half of $\lambda_{SP}$. One has to conclude that the distances from the NFO probe moduli $\lambda_{SP}/2$ are the determining parameters for the undulation of SP excitation. This is suggestive of (i) reflection of the tip-generated SP by the protrusions, which will result in SP interferences, and (ii) an influence of the SP amplitude at the tip position on the coupling process.

To confirm this hypothesis, a numerical simulation based on the coupled dipole method [25] was performed [26]. Using Green's functions of the layered reference system, the protrusions as well as the optical probe were modeled as interacting dipolar centers. The resulting scan images — an example is depicted in Fig. 6b — bear striking similarity to the experimental ones. Note for instance the zones of reduced modulation amplitude that might be caused by interference of several reflected SP waves. Another interesting feature is the break-up of the circles into two slightly displaced half-arcs near the scattering centers — a feature that apparently is correlated to the dipole character of the scattering centers. Although the underlying physical processes have not yet been completely clarified, it is clear that the amplitude of undulation must be proportional to the *SP/SP* scattering efficiency of an imperfection within the film in addition to factors that depend on the distance between NFO probe and scattering center.

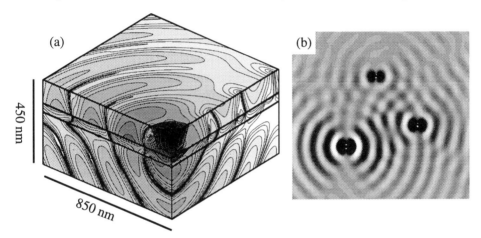

*Figure 6.* *Computer simulations*: (a) SP excitation in a 50-nm-thick gold film ($\epsilon = -3.47 + I2.84$ ) on a glass substrate by an aperture-SNOM probe with aluminum cladding ($\epsilon = -38.5 + I10.2$), $\lambda_0 = 515$ nm, snapshot for a given arbitrary phase. Contours: electrical energy density, factor of two between adjacent lines. Note preferential propagation into direction of polarization (left), the longitudinal character of the wave inside the film, and the radiation into the direction of $\theta_{SP}$. (b) Forbidden-light scan image of the silver film with three dipoles representing perturbations (reminder: this is a *computer simulation!*).

The weak radiation giving rise to the allowed image (Fig. 5c) may contain directly transmitted light, which could carry information about the *local* optical properties of the film. The undulation pattern, however, cannot have this origin. We presume that it is produced by SP scattering at the imperfections that can emit radiation into all directions. The formation of hyperbolae is suggestive of interactions for which the *difference* in distance from the NFO probe modulo $\lambda_{SP}/2$ is the dominant parameter. This is the case for the *relative phase* of the SP waves at two neighboring protrusions. The radiation emitted from these centers will carry the same phase differences. The contributions from the different centers emitted into the allowed directions interfere at the photocathode of the far-away detector. The path difference outside the film is negligibly small as required for the explanation of the observed hyperbolic features. The amplitude of interference modulation must hence be proportional to the *out-of-film, SP/photon* scattering efficiency of the protrusions.

A close look at the protrusions in Figs. 5b and 5c shows that the centers of the undulation patterns all are slightly displaced to the right with respect to the sharply defined bright and dark spots that coincide exactly with those seen in the allowed and forbidden topographic images, respectively. We therefore suspect that the bright spots are artifacts created by the distance dependence of the coupling efficiency [27] rather than real NFO features. The displacement can then be readily understood as the distance between an SFD-active protruding crystallite of the aluminum coating and the center of the aperture as sketched in Figs. 2c and 2d.

## 5. Summary

Angular distributions and approach curves clearly demonstrate the excitation of SP with an NFO probe. The three scan images shown provide a unique combination of data about the imperfections of the silver film: topography, SP scattering efficiency within the film, and emission efficiency due to scattering out of the film. This may open a new way to systematically characterize the properties of structured metal films, as the theoretical tools needed for interpretation have also been made available. Optimization of these structures is of practical interest for applications such as adsorbate spectroscopy, Raman scattering or the generation of surfaces with special color and luster.

We thank H. Bielefeldt, G. Eggers, Ch. Hafner, H. Heinzelmann and B. Hug for helpful discussions and collaboration.

## References

1. Raether, H. (1988) *Surface Plasmons on Smooth and Rough Surfaces and on Gratings*, volume 111 of *Springer Tracts in Modern Physics*. Springer Verlag, Berlin Heidelberg.
2. Fischer, U.Ch. and Pohl, D.W. (1989) Observation on single-particle plasmons by near-field optical microscopy, *Phys. Rev. Lett.* **62**, 458.
3. Specht, M., Pedarnig, J.D., Heckl, W.M. and Hänsch, T.W. (1992) Scanning plasmon near-field microscope. *Phys. Rev. Lett.* **68**, 476.

4.  Rücker, M., Knoll, W. and Rabe, J. (1992) Surface-plasmon-induced contrast in scanning tunneling microscopy. *J. Appl. Phys.* **72**, 5027.
5.  Kim, Y.-K., Lundquist, P.M., Helfrich, J.A., Mikrut, J.M., Wong, G.K. and Auvil, P.R. (1995) Scanning plasmon optical microscope. *Appl. Phys. Lett.* **66**, 3407.
6.  Marti, O., Bielefeldt, H., Hecht, B., Herminghaus, S., Leiderer, P. and Mlynek, J. (1993) Near-field optical measurement of the surface plasmon field. *Optics Commun.* **96**, 225.
7.  Dawson, P., de Fornel, F. and Goudonnet, J.-P. (1994) Imaging of surface plasmon propagation and edge interaction using a photon scanning tunneling microscope. *Phys. Rev. Lett.* **72**, 2927.
8.  de Hollander, R.B.G., van Hulst, N.F. and Kooyman, R.P.H. (1995) Near field plasmon and force microscopy. *Ultramicroscopy* **57**, 263.
9.  Tsai, D.P., Kovacs, J., Moskovits, M., Shalaev, V. and Suh, J.S. (1994) Photon scanning tunneling microscopy of optical excitations of fractal metal colloid clusters. *Phys. Rev. Lett.* **72**, 4149.
10. Bozhevolnyi, S.I., Vohnsen, B., Smolyaninov, I. and Zayats, A.V. (1995) Direct observation of surface polariton localization caused by surface roughness. *Optics Commun.* **117**, 417.
11. Dürig, U., Pohl, D.W. and Rohner, F. (1986) Near-field optical-scanning microscopy. *J. Appl. Phys.* **59**, 3318.
12. Novotny, L., Pohl, D.W. and Hecht, B. (1995) Scanning near-field optical probe with ultrasmall spot size. *Opt. Lett.* **20**, 970.
13. Kretschmann, E. and Raether, H. (1968) Radiative decay of non-radiative surface plasmons excited by light. *Z. Naturforsch.* **23a**, 2135.
14. Otto, A. (1968) Excitation of nonradiative surface plasma waves in silver by the method of frustrated total reflection. *Z. Phys.* **216**, 398.
15. Hecht, B., Heinzelmann, H. and Pohl, D.W. (1995) Combined aperture SNOM/PSTM: Best of both worlds? *Ultramicroscopy* **57**, 228.
16. Hecht, B., Heinzelmann, H., Novotny, L. and Pohl, D.W. (1995) 'Tunnel' near-field optical microscopy: TNOM-2. In O. Marti and R. Möller, Eds., *Photons and Local Probes,* NATO ASI Series E: Applied Sciences, Vol. 300, pp. 93–108. Kluwer Academic Publishers, Dordrecht.
17. Pohl, D.W., Denk, W. and Lanz, M. (1984) Optical stethoscopy: Image recording with resolution $\lambda/20$. *Appl. Phys. Lett.* **44**, 651.
18. Betzig, E., Isaacson, M. and Lewis, A. (1987) Collection mode near-field scanning optical microscopy. *Appl. Phys. Lett.* **51**, 2088.
19. Courjon, D., Sarayeddine, K., and Spajer, M. (1989) Scanning tunneling optical microscopy. *Optics Commun.* **71**, 23.
20. Reddick, R.C., Warmack, R.J., and Ferrell, T.L. (1989) New form of scanning optical microscopy. *Phys. Rev. B* **39**, 767.
21. Lefevre, H.C. (1980) Single-mode fiber fractional wave devices and polarization controller. *Electronics Lett.* **16**, 778.
22. Rugar, D., Mamin, H.J., Erlandson, R., Stern, J.E. and Terris, B.D. (1988) Microscope using a fiber-optic displacement sensor. *Rev. Sci. Instrum.* **59**, 2337.
23. Novotny, L. and Pohl, D.W. (1995) Light propagation in scanning-near field optical microscopy. In O. Marti and R. Möller, Eds., *Photons and Local Probes,* NATO ASI Series E: Applied Sciences, Vol. 300, pp. 21–33. Kluwer Academic Publishers, Dordrecht.
24. Hafner, Ch. (1990) *The Generalized Multiple Multipole Technique for Computational Electromagnetics.* Artech, Boston.
25. Taubenblatt, M.A. and Tran, T.K. (1993) Calculation of light scattering from particles and structures on a surface by the coupled-dipole method. *J. Opt. Soc. Am. A* **10**, 912.
26. Novotny, L., *et al.* (in preparation).
27. Hecht, B., *et al.* (in preparation).

# WEAK LOCALIZATION OF SURFACE PLASMON POLARITONS: DIRECT OBSERVATION WITH PHOTON SCANNING TUNNELING MICROSCOPE

S.I. BOZHEVOLNYI, A.V. ZAYATS[*] and B. VOHNSEN
*Institute of Physics, Aalborg University,*
*Pontoppidanstræde 103, DK-9220 Aalborg, Denmark*
*[*]Institute of Spectroscopy, Russian Academy of Sciences,*
*142092, Troitsk, Russia*

ABSTRACT. Using a photon scanning tunneling microscope with shear force feedback we directly probe optical fields of surface plasmon polaritons (SPPs) excited at surfaces of two silver films with different roughness. We observe that near-field optical images exhibit interference between the excited and multiple scattered SPPs. The orientation and period of interference fringes observed relatively far from surface scatterers indicate the presence of the backscattered SPP. These fringes are found to be more pronounced for the silver film with larger roughness. We relate the observed phenomenon to weak localization of SPPs caused by multiple SPP scattering in the surface plane.

## 1. Introduction

Scattering of surface plasmon polaritons (SPPs) by surface roughness is one of the subjects, which study can be significantly complemented by using experimental techniques developed in scanning near-field optical microscopy. These techniques open a new possibility for investigations of SPP properties virtually at the surface studied. The choice of the appropriate technique depends on the problem under consideration. Photon scanning tunneling microscope (PSTM) with an uncoated fiber tip is apparently the most suitable technique for local probing of the SPP field, since such a tip can be within certain approximations considered as a nonperturbative probe of the electric field intensity [1]. The PSTM has been already used to measure the degree of SPP field enhancement [2] and the decay of the SPP evanescent field as a function of the tip-surface distance [3], as well as to study the SPP propagation along

163

*M. Nieto-Vesperinas and N. García (eds.), Optics at the Nanometer Scale 163–173.*

thin silver films [4] and the SPP field behaviour in fractal colloid clusters [5]. The main drawback in the aforementioned studies is related to the fact that the surface profile could not have been measured along with the optical signal. It is well known that, without knowledge of the surface profile, the interpretation of optical images is cumbersome and can be ambiguous [6].

We have recently constructed the PSTM with the shear force based feedback system and studied optical fields of SPPs excited at different interfaces of gold films while imaging simultaneously surface relief structures [7,8]. We observed that near-field optical images, which are generated due to the SPP propagating along a relatively smooth surface, show a well pronounced interference pattern related to the interference between the excited and scattered SPPs [7], whereas those due to the SPP at a rough surface exhibit spatially localized (within 150 - 250 nm) field enhancement by up to 7 times [8]. The latter phenomenon has been recognized as the strong (Anderson) localization of SPPs in the plane of the sample surface [7,8].

In order to understand the observed difference in the regimes of SPP scattering one should relate the SPP propagation length $L$ to the distance $l$ between surface scatterers. For the gold films used and the wavelength of light $\lambda \approx 633$ nm, the propagation length can be estimated as $L \approx 8.5$ $\mu$m by using the measured angular dependencies of the reflected light power [7]. We have recorded the topographical images of various regions of the smooth gold film and concluded that, for this film, the distance $l$ is of the same magnitude as the propagation length $L$. This explains the fact that the appropriate near-field optical images exhibit interference patterns, which are typical for the regime of *single* scattering [7]. For the rough gold film, the topographical images show clearly that the distance $l$ is on the submicron scale [7,8], i.e. $L \gg l$. Therefore, the regime of *multiple* scattering of SPPs is developed in this case leading to the (strong) SPP localization [7,8]. Apparently, there can be realized an intermediate situation with $L > l$, which would result in enhanced backscattering (or weak localization) of SPPs. Note, that in all cases it is very important that the scattering of SPPs into radiative modes (out of the surface plane) should be sufficiently weak.

Enhanced backscattering of light from random media (including rough surfaces) has been studied both theoretically and experimentally for the last ten years (see [9-11] and references therein). Recently, backscattering enhancement arising from the SPP excitation on a rough metal surface has been observed in the diffusely scattered light [10]. This enhancement has been found to result from the coherent interference of scattered contributions (propagating in air) from counterpropagating SPPs [11]. However, enhanced backscattering of SPPs *in the surface plane* has been observed only indirectly by detecting a sharp peak in the angular dependence of the efficiency of second harmonic generation

(SHG) in the direction perpendicular to the sample surface [12]. The occurrence of such a peak is a fingerprint of the enhanced backscattering of SPPs, since SHG in the normal direction is related to the nonlinear interaction between counterpropagating (i.e. between the excited and backscattered) SPPs at the fundamental frequency [13]. In the following, the phenomenon of enhanced backscattering of SPPs in the surface plane will be referred to as weak localization of SPPs in order to distinguish it from the backscattering enhancement in the diffusely scattered (out of the surface plane) light [10,11].

Here, we present results of an experimental study of scattering of SPPs excited at weakly rough surfaces of silver films analogous to those used in the SHG experiments [12]. Optical near-field and topographical images are simultaneously obtained by using the PSTM with shear force regulation of the tip-surface distance [7,8]. We observe that the optical images exhibit interference between the excited and multiple scattered SPPs. The orientation and period of interference fringes indicate the presence of the backscattered SPP. We relate the observed phenomenon to the weak localization of SPPs.

## 2. Experimental technique

The experimental setup, which consists of the PSTM combined with the shear force based feedback system and an arrangement for SPP excitation in the usual Kretschmann configuration, is described in detail elsewhere [7]. The p-polarized (electrical field is parallel to the plane of incidence) light beam from a He-Ne laser ($\lambda \approx 633$ nm, $P \approx 3$ mW) is used for the SPP excitation, which is recognized as a minimum in the angular dependence of the reflected light power. We studied excitation and scattering of SPPs on two slightly different silver films (but with the same thickness of 45 nm) fabricated under conditions similar to those used when preparing the samples for the SHG experiments [12]. The first film has been thermally evaporated with the speed of $\sim 1$ Å/s in vacuum ($\approx 10^{-6}$ Torr). In order to obtain a film with larger surface roughness, the second film has been evaporated at a greater speed ($\sim 5$ Å/s) in a relatively poor vacuum ($\approx 5 \cdot 10^{-5}$ Torr). We have used commercially available optical glass prisms (refractive index $n = 1.44$) as the substrates in order to facilitate the SPP excitation in the Kretschmann configuration. Angular dependencies of the reflected light power were measured for both films in order to determine the excitation angles for the appropriate SPPs (Fig. 1). It is seen that both dependencies have essentially the same behaviour and exhibit well pronounced minima at the angle $\theta \approx 46^0$ of light incidence with the angular width of $\approx 2^0$. By fitting the experimentally obtained angular dependencies to those calculated for the appropriate layered structure, the SPP wavelength $\lambda_{SPP}$ and propagation length $L$ have been found to be $\lambda_{SPP} \approx 610$ nm and $L \approx 25$ $\mu$m for both films.

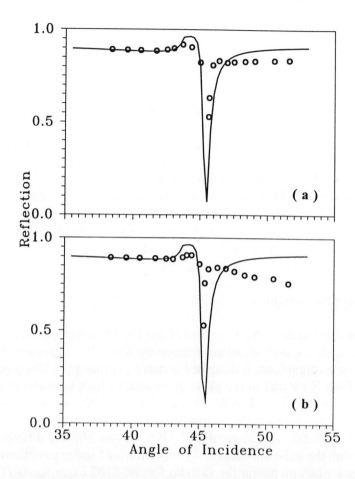

*Figure 1.* Angular dependencies of the reflected (p-polarized) light power measured (circles) and calculated (solid lines) for the 45-nm-thick silver films: (a) - film #1; (b) - film #2.

The SPP local field is probed with an uncoated fiber tip, which is fabricated by etching of a single-mode silica fiber in a 40% solution of hydrofluoric acid during a time period of 55 min [14]. Similar fiber tips have been used in our previous studies of SPP propagation and scattering on gold films [7,8]. These tips have a rather large cone angle (typically $\delta \approx 40^0$), which is of crucial importance in our experiments for the following reasons. The SPP field is predominantly polarized in the direction perpendicular to the surface plane, and the detected signal related to the perpendicular field component increases rapidly with the cone angle ($\sim \delta^4$, for small $\delta$) [15]. For example, we have found that the SPP related signal measured with a sharper fiber tip ($\delta \approx 10^0$) was about two orders of magnitude weaker than the signal detected with the tip that we usually used [7,8]. Discrimination of the propagating waves, which result from the SPP scattering out of the surface plane, against the evanescent field of the excited and scattered (in the surface plane) SPPs is also much more efficient with such a tip. The propagating (in air) field contains components parallel to the surface plane, and the detected signal related to these components increases with the cone angle as $\delta^2$ (for small $\delta$) [15]. Consequently, the relative contribution of the perpendicular (to the surface plane) field components in the detected signal increases with the cone angle implying the increase of the SPP related signal. For example, the detected signal was typically more than 20 times smaller if the fiber tip was moved ~ 1 $\mu$m away from the surface of a smooth metal (gold [7] or silver) film with the SPP being resonantly excited. Under similar circumstances but with the aforementioned sharp tip, we have found only a 60% decrease in the detected signal. Therefore, the usage of the fiber tip with a large cone angle is essential for our experiments, in which the main purpose is to probe the fields of the excited and scattered (in the surface plane) SPPs. Finally, it should be noted that all images presented here are oriented in the way that the excited SPP propagates upwards in the vertical direction.

## 3. SPP scattering

Topographical images of the silver film surface of the first sample showed a smooth surface with rarely spaced submicron-sized bumps with 40-80 nm heights (Fig. 2a). The average distance $l$ between these scatterers was estimated to be about 8 $\mu$m. Excitation of the SPP at the silver-air interface of this sample exhibited a well pronounced resonance behaviour, which was expected from the far-field measurements (Fig. 1), and the detected (near-field) optical signal was up to 5 nW for the resonant excitation. The average optical signal was about 25 times smaller if the angle $\theta$ of beam incidence on the film surface was out of resonance by $\approx 2^0$ or if the fiber tip was moved a few microns away from the surface. Therefore, we can disregard the contribution of propagating in air components of the scattered field in the detected signal. If a scatterer was present in the field of view of our microscope, the near-field optical image (with the SPP being resonantly excited) showed an interference pattern related to the interference between the excited and scattered SPPs (Fig. 2b). We have observed similar (but more pronounced) interference patterns in our previous experiments with the smooth gold film [7]. However, careful examination of the interference pattern recorded with the silver film reveals also the presence of SPPs scattered from other scatterers than the one seen on the topographical image (Fig. 2). This difference in the optical images of gold and silver films, which have similar topography (compare this Fig. 2 and Fig. 7 from ref. [7]), is related to the circumstance that the SPP propagation length $L$ for the silver film is about 3 times larger than that for the gold film. Consequently, $L > l$ for the silver film (whereas $L \sim l$ for the gold film), and one can expect to observe interference effects related to the regime of multiple scattering of SPPs propagating along the silver film surface (see Section 1).

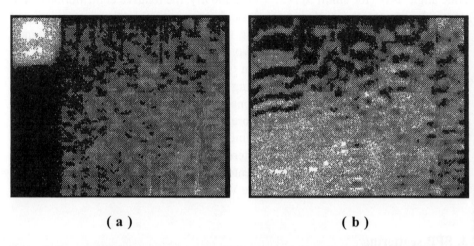

( a )                                          ( b )

*Figure 2.* Gray-scale topographical (a) and near-field optical (b) images 5×4.5 $\mu m^2$ obtained with the first silver film. The maximum depth of the topographical image is 73 nm. The scale of the optical image corresponds to ~ 0 - 2 nW of the detected optical signal.

Weak localization of SPPs should manifest itself in the formation of the backscattered SPP due to interference effects in multiple scattering of SPPs in the surface plane. The physical origin of this two-dimensional phenomenon is evidently the same as that of the enhanced backscattering in three dimensions: the waves, which travel along the same light path in opposite directions, will *always* have the same phase and interfere constructively *only* in the direction of pure backscattering [9]. Apparently, formation of the backscattered SPP should be most pronounced relatively far away from individual scatterers, where SPPs scattered along different routes are of the same magnitude. In three dimensions, this situation corresponds to the detection of scattered waves outside of a medium with randomly placed scatterers [9]. The presence of the backscattered SPP can be then deduced from, in general, a complicated interference pattern by extracting the *specific* pattern corresponding to the interference between the excited and backscattered SPPs. Since the excited SPP propagates upwards in our images, the appearance of horizontally oriented fringes with the period equal to half of the SPP wavelength $\lambda_{SPP}$ on the near-field optical image can be regarded as an evidence of the presence of the backscattered SPP.

We recorded several topographical and near-field optical images in different places of the first film, which exhibited a smooth surface relief (with the maximum depth being less than 20 nm) without well defined scatterers (Fig. 3a). The contrast of the optical images was usually low, so that the interference fringes could be hardly seen on these images. However, after averaging and filtering low spatial frequencies out of the optical images, horizontally oriented fringes showed up quite distinctly (Fig. 3b). The period of these fringes corresponds to the expected value of $\lambda_{SPP}/2 \approx 305$ nm within the accuracy of 5% of our measurements, thereby indicating the presence of the backscattered SPP.

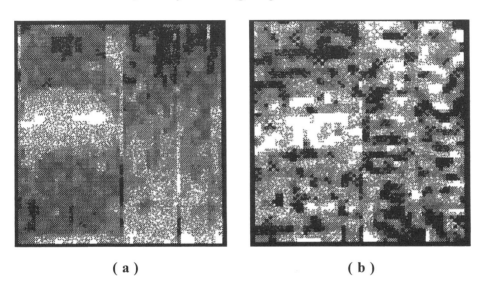

**( a )**　　　　　　　　　　　　**( b )**

*Figure 3.* Gray-scale topographical (a) and near-field optical (b) images 3×3 $\mu m^2$ obtained with the first silver film. The maximum depth of the topographical image is 18 nm. The optical image is processed by averaging and filtering low spatial frequencies out.

It is clear that the horizontal fringes are just hidden in the unprocessed optical image (Fig. 4a) and can be either visualized by using the appropriate image processing (Fig. 3b) or found by performing the Fourier transformation of the image. The Fourier transform of the unprocessed optical image (Fig. 4a) containing 64×64 pixels was calculated by using the standard Fast Fourier Transform (FFT) routine (Fig. 4b). Note, that the central part of the FFT image, which corresponds to low spatial frequencies (i.e. to an average level of the detected optical signal and its slow variations), is not shown in Fig. 4b. Bright spots along the central horizontal line of the FFT image correspond to the signal variations between successive scans (scanning direction is vertical in our images). These variations are topographically induced because of somewhat instable behaviour of the shear force feedback with this sample (Fig. 3a), and,

170

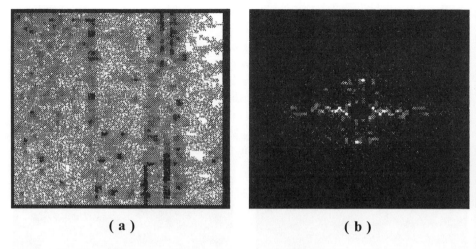

( a )                                                    ( b )

*Figure 4.* The unprocessed near-field optical ~ 3×3 $\mu$m$^2$ image (a) corresponding to the processed image in Fig. 3b and its Fourier transform (b) with the central part being omitted. The scale of the optical image corresponds to ~ 0.5 - 2 nW of the detected optical signal.

therefore, of no importance for the present discussion. However, the FFT image (Fig. 4b) exhibit clearly two very significant bright spots situated symmetrically along the central vertical line. These spots correspond to the strictly periodic variation in the detected optical signal with the period of $\lambda_{SPP}/2 \approx 305$ nm. It means that the horizontal fringes seen on the processed optical image (Fig. 3b) are not an artifact related to the specific image processing, and that the presence of the backscattered SPP is indeed demonstrated with the optical image obtained.

Topographical images of the silver film surface of the second sample showed that the surface topography of the second film is similar to that of the first film (compare Figs. 2a and 5a) but with more closely situated surface bumps. The average distance $l$ between the scatterers was estimated to be about 5 $\mu$m, which is noticeably less than the value of $l$ for the first film. Actually, the decrease in $l$ has been expected as a consequence of the difference in the fabrication procedures for these films (see Section 2). Excitation of the SPP at the silver-air interface of this sample exhibited a resonance behaviour similar to that observed with the first film. However, the near-field optical signal (for the resonant SPP excitation) was about 5 times less than the signal detected with the first film, which is probably connected with the increase of the SPP scattering. The average optical signal was about 20 times smaller if the angle $\theta$ of beam incidence on the film surface was out of resonance by $\approx 2^0$ or if the fiber tip was moved a few microns away from the surface. Therefore, with this film we can also disregard the contribution of propagating in air components of the scattered field in the detected signal.

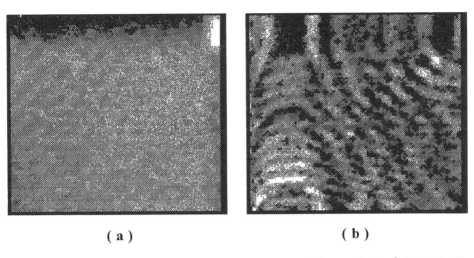

**( a )**   **( b )**

*Figure 5.* Gray-scale topographical (a) and near-field optical (b) images 5×5 $\mu m^2$ obtained with the second silver film. The maximum depth of the topographical image is 57 nm. The scale of the optical image corresponds to ~ 0 - 0.2 nW of the detected optical signal.

The near-field optical images obtained with the second sample were similar to those recorded with the first sample, but the interference effects related to multiple scattering of SPPs were much more pronounced. Even in the presence of scatterer in the field of view, the optical image showed a complicated interference pattern indicating the presence of several scattered SPPs (Fig. 5b). On this image, one can already see appearance of the characteristic horizontal fringes associated with the presence of the backscattered SPP. The difference in the optical images of two surface regions with similar topography but of different films (cf. Figs. 2 and 5) is apparently related to the difference in the average distance $l$ between surface scatterers for these films: multiple scattering of SPPs is more developed in the case of smaller $l$ (i.e. with the second film) resulting in more pronounced interference effects. One should expect to find the similar enhancement of contrast in the optical images recorded in the absence of scatterers in the field of view. It was quite difficult to locate such a place on the surface of the second film because of the relatively small value of $l$, but the results obtained were rewarding: the optical image exhibited the appropriate horizontal interference fringes seen clearly without any image processing (Fig. 6). Even though the difference between the optical images recorded in the absence of scatterers with the studied films (Figs. 4a and 6b) is remarkable, the processed optical image of the first sample (Fig. 3b) is rather similar to the unprocessed one of the second sample (Fig. 6b). This demonstrates once again that the essential content of these images is the same, namely, the horizontal interference fringes due to the presence of the backscattered SPP.

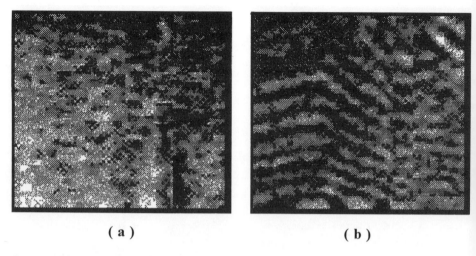

( a )　　　　　　　　　　　　　　　　( b )

*Figure 6.* Gray-scale topographical (a) and near-field optical (b) images 3.5×3.5 $\mu m^2$ obtained with the second silver film. The maximum depth of the topographical image is 16 nm. The scale of the optical image corresponds to ~ 0 - 0.2 nW of the detected optical signal.

## 4. Conclusions

Using the PSTM with shear force feedback we have studied the SPP scattering (in the surface plane) with two 45-nm-thick silver films fabricated under different conditions, which resulted in different densities of surface scatterers. The SPP excitation at silver-air interfaces of these films has been investigated with far- and near-field optical techniques. We have found that the propagating in air components of the scattered field can be disregarded in near-field optical measurements, if the SPP is resonantly excited and detected with the appropriate fiber tip (with a large cone angle). We have observed that the near-field optical images exhibit interference between the excited and multiple scattered SPPs. The occurrence of multiple scattering of SPPs in the surface plane has been explained by the circumstance that the SPP propagation length was larger than the average distance between surface scatterers. It has been found that, relatively far from surface scatterers, interference fringes are oriented horizontally and have the period equal to half of the SPP wavelength, thereby indicating the presence of the backscattered SPP. These fringes have appeared more pronounced and clearly visible for the film with larger roughness. It should be mentioned, that, in the aforementioned SHG experiments [12] with the silver films analogous to those used in the present work, the efficiency of SHG in the direction perpendicular to the surface has been found to be significantly larger for films with larger roughness. Overall, we conclude that the results obtained demonstrate weak localization of SPPs due to interference effects in multiple SPP scattering in the surface plane of weakly rough silver films.

# 5. References

1. Carminati, R. and Greffet, J.-J. (1995) Two-dimensional numerical simulation of the photon scanning tunneling microscope. Concept of transfer function, *Opt. Commun.* **116**, 316-321.
2. Bielefeldt, H., Hecht, B., Herminghaus, S., Mlynek, J., and Marti, O. (1993) Direct measurement of the field enhancement caused by surface plasmons with the scanning tunneling optical microscope, in D.W. Pohl and D. Courjon (eds.), *Near Field Optics*, Kluwer Academic Publishers, Dordrecht, pp. 281-286.
3. Adam, P.M., Salomon, L., de Fornel, F., and Goudonnet, J.P. (1993) Determination of the spatial extension of the surface-plasmon evanescent field of a silver film with a photon scanning tunneling microscope, *Phys. Rev. B* **48**, 2680-2683.
4. Dawson, P., de Fornel, F., and Goudonnet, J.P. (1994) Imaging of surface plasmon propagation and edge interaction using a photon scanning tunneling microscope, *Phys. Rev. Lett.* **72**, 2927-2930.
5. Tsai, D.P., Kovacs, J., Wang, Z., Moskovits, M., Shalaev, V.M., Suh, J.S., and Botet, R. (1994) Photon scanning tunneling microscopy images of optical excitations of fractal metal colloid clusters, *Phys. Rev. Lett.* **72**, 4149-4152.
6. van Hulst, N.F., Segerink, F.B., Achten, F., and Bölger, B. (1992) Evanescent-field optical microscopy: effects of polarization, tip shape and radiative waves, *Ultramicroscopy* **42**, 416-421.
7. Bozhevolnyi, S.I., Smolyaninov, I.I., and Zayats, A.V. (1995) Near-field microscopy of surface-plasmon polaritons: localization and internal interface imaging, *Phys. Rev. B* **51**, 17916-17924.
8. Bozhevolnyi, S.I., Vohnsen, B., Smolyaninov, I.I., and Zayats, A.V. (1995) Direct observation of surface polariton localization caused by surface roughness, *Opt. Commun.* **117**, 417-423.
9. van Albada, M.P., van der Mark, M.B., and Lagendijk, A. (1990) Experiments on weak localization of light and their interpretation, in P. Sheng (ed.), *Scattering and Localization of Classical Waves in Random Media*, World Scientific, Singapore, pp. 97-136.
10. West, C.S. and O'Donnel, K.A. (1995) Observations of backscattering enhancement from polaritons on a rough metal surface, *J. Opt. Soc. Am. A* **12**, 390-397.
11. Michel, T.R., Knotts, M.E., and O'Donnel, K.A. (1995) Scattering by plasmon polaritons on a rough surface with a periodic component, *J. Opt. Soc. Am. A* **12**, 548-559.
12. Aktsipetrov, O.A., Golovkina, V.N., Kapusta, O.I., Leskova, T.A., and Novikova, N.N. (1992) Anderson localization effects in the second harmonic generation at a weakly rough metal surface, *Phys. Lett. A* **170**, 231-234.
13. McGurn, A.R., Leskova, T.A., and Agranovich, V.M. (1991) Weak-localization effects in the generation of second harmonics of light at a randomly rough vacuum-metal gratings, *Phys. Rev. B* **44**, 11441-11456.
14. Bozhevolnyi, S.I., Keller, O., and Xiao, M. (1993) Control of the tip-surface distance in near-field optical microscopy, *Appl. Opt.* **32**, 4864-4868.
15. Van Labeke, D. and Barchiesi, D. (1993) probes for scanning tunneling optical microscopy: a theoretical comparison, *J. Opt. Soc. Am. A* **10**, 2193-2201.

# STM-INDUCED PHOTON EMISSION FROM AU(110)

*The Role of the Topography*

RICHARD BERNDT

*Université de Lausanne*
*Institut de Physique Expérimentale*
*CH-1015 Lausanne, Switzerland*

**Abstract.** A low-temperature ultra-high vacuum scanning tunneling microscope (STM) is used to excite photon emission from Au(110) surfaces. In the detected photon intensity the (1×2) reconstruction of the Au surface is clearly resolved. This observation is interpreted in terms of local variations of the electromagnetic interaction of tip and sample occurring at constant tunneling current.

## 1. Introduction

During the last decade a number of scanning probe microscopies have been developed which measure various sample properties with the help of the diverse interactions that exist between a tip and a sample [1]. Most of these microscopies are performed under the condition of constant tip-sample distance and an independent distance regulation based on e.g. a scanning tunneling microscope (STM) or a kind of force microscopy is used to define and maintain this distance. As a consequence, the isolation of topography effects in the measured data is an important issue. Below we will present results of a particular scanning probe method, STM-induced photon emission [2], for clean Au(110) surfaces. On these surfaces, apparent atomic resolution in the photon emission was observed [3]. An analysis reveals that this resolution is due to the variation of the electromagnetic coupling of the tip and the Au sample on an atomic scale. This variation results from the vertical motion of the STM tip which is caused by the atomic structure of the surface that is resolved in the constant-current STM image.

*M. Nieto-Vesperinas and N. García (eds.), Optics at the Nanometer Scale 175–180.*

## 2. Experiment

The experiments were performed with a custom-built ultra-high vacuum STM employing low temperatures (5K and 50K) [4]. The Au(110) and Ag(110) surfaces and the W tips were prepared by standard procedures consisting of Ne or Ar ion bombardment and heating in ultra-high vacuum. Photons emitted from the tunneling gap were detected with a photomultiplier which was sensitive to the wavelength range 200nm $< \lambda <$ 800nm. In order to measure *photon maps* of the surface the detected photon count rate was recorded by the STM electronics quasi-simultaneously with constant-current topographs for each image pixel.

## 3. Results

### 3.1. DISTANCE DEPENDENCE OF THE PHOTON EMISSION

The sub-nanometer proximity of the tip and the sample of the STM gives rise to electromagnetic coupling of tip and sample [5, 6]. As a consequence, new localized plasmons (tip-induced plasmons — TIP) are induced which are specific to the STM geometry. These TIP modes are excited via inelastic tunneling processes and can decay radiatively giving rise to photon emission. A consequence of this interpretation is that the photon emission yield is expected to vary with the distance between tip and sample.

**Figure 1.** *Yield of STM-induced photon emission (defined as photon intensity divided by tunneling current) as a function of tip displacement $\Delta s$ observed from Cu(111) using a W tip. The response of the detector was limited using a band-pass filter centered at $\lambda = 600nm$. The experimental data (dots) agrees nicely with the result of a model calculation (solid line). Adapted from Ref. [6].*

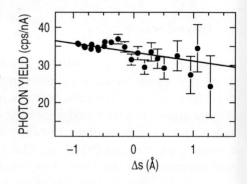

This variation was measured in Ref. [6] (Fig. 1) by changing the tip-sample distance in a controlled manner. A vertical displacement $\Delta s$ of the tip causes a variation of the tunneling current $I_t$ and of the photon intensity. From simultaneous measurements of $I_t$ and the intensity of photons with $\lambda \approx 600nm$ as a function of tip-sample distance the photon yield was defined as the ratio of intensity to current. In Figure 1, a decrease in photon yield (dots) with increasing distance is discernible over the limited range of tip excursions studied. The experimental data is reproduced quite well by a model calculation (solid line). This calculation revealed that the observed magnitude of the yield decrease results from two counteracting effects. First, the decrease of the field strength of the TIP modes in the tunneling gap was predicted to result in a larger drop of the emitted intensity ($\approx 30\%/\text{Å}$) than in Fig. 1 ($\approx 7\%/\text{Å}$). Next, this effect is partially compensated by the longer tunneling distance between tip and sample which enters the calculation through the transition matrix elements between initial and final states.

While the data in Fig. 1 was measured on a Cu(111) surface a similar result is expected for Au(110) within the model of TIP modes as well. An important finding is that the above data show a detectable variation of the strength of the TIP modes caused by sub-Angstrom changes of the tip-sample distance. An exact solution of the Maxwell equations for a tip-surface geometry has recently confirmed this sensitive dependency of the TIP modes on the tip-sample separation (see the chapter by Madrazo, Nieto-Vesperinas and Garcia in these proceedings).

## 3.2. ATOMIC RESOLUTION

Figure 2a displays a STM image of a Au(110) surface which is comprised of terraces of close-packed gold rows and monoatomic steps.

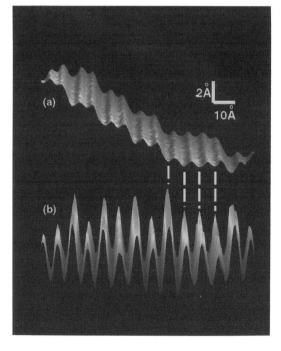

**Figure 2.** *Constant current STM image (a) and photon map (b) of a stepped area of a (1×2) reconstructed Au(110) surface measured simultaneously. In (b) the photon intensity is represented by height and grey level. Close-packed gold rows which are separated by 8.16Å are resolved in the topograph. An identical but phase shifted periodicity is found in the photon map. The dashed lines serve to guide the eye. Increased photon intensity maxima are observed near steps. The overall modulation of the photon intensity is $\approx \pm 25\%$ of the average intensity. Temperature of tip and sample: 50K, $I_t = 4.4nA$, $V_t = 3.0V$. From Ref. [3]*

The distance between the atomic rows is $\approx 8.2\text{Å}$. A similar periodic variation is resolved in the photon intensity map recorded simultaneously (Fig. 2b). As indicated by dashed lines higher photon intensities are observed when the tip is placed between two atomic rows. Moreover, the intensity modulation is distinctly enhanced near steps.

The contrasts in photon maps from metal surfaces result from local variations in the inelastic tunneling current. Possible origins of such variations encompass local variations of the density of final states for inelastic processes and of the tunneling barrier. As discussed in detail in Ref. [3] these mechanisms can be ruled out in the present case.

An interpretation of the atomic resolution is suggested by the data of Fig. 1. The electromagnetic field strength of the TIP modes and, thus, the photon yield depend sensitively on the distance between tip and sample. Owing to the lateral extent of the TIP modes which has been estimated theoretically to be of the order of 50Å [8] the distance which is relevant for the photon emission yield is expected to represent a lateral average over a surface area of similar dimensions. This degree of lateral localization should be compared to the lateral extent of the tunneling current of a few Angstroms [9]. Therefore, the vertical movement of the tip which is caused by the constant-current distance regulation gives rise to a modulation of the photon intensity. This effect is consistent with the observed contrasts in photon maps. First, the larger distance between the tip and the *averaged* surface which occurs at topographic maxima (atomic rows) weakens the TIP modes and hence decreases the photon emission in agreement with the experimental observation. Next, the expected size of this effect is not far from the experimental value of Fig. 2b: the predicted variation of the photon-intensity with tip-sample distance (40%/Å) and the measured topographic corrugation in Fig. 2a (0.7Å) suggest a $\approx 30\%$ intensity modulation.

The distinct intensity variations observed at steps (Fig. 3) provide a more serious test of the above outlined concept. In order to estimate the photon emission intensity from a corrugated surface $h(x)$ we define an average surface height from a convolution of the topographic surface with the lateral extent of the TIP mode. For sub-Angstrom deviations of the actual tip position (i.e. the constant-current STM image) from this averaged surface, a linear term is sufficient to describe the resulting intensity variations. Thus, we calculate the photon emission $p(x)$ from a corrugated surface to be:

$$p(x) = p_0 - \alpha \left( h(x) - \int_{-\infty}^{\infty} h(\tilde{x})\, g(x - \tilde{x})\, d\tilde{x} \right),$$

where $p_0$ is the average photon intensity and $\alpha$ is a constant. $g(x)$ is a weighting function representing an approximately Gaussian variation of the electrical field strength with lateral distance from the tip axis.

**Figure 3.** *Comparison of cross-sections through (a) an STM image of a stepped area of a Au surface, (b) the corresponding measured photon map, and (c) the result of a model calculation for this surface topography. In the topographic data (a), atomic rows and several small terraces separated by monoatomic steps are resolved. The photon intensity (b) oscillates with the atomic periodicity. The photon intensity maxima are shifted with respect to the topograph and significantly stronger photon intensity maxima occur when the STM tip is placed at the lower side of steps. The model calculation (c) reproduces the essential features of (b).*

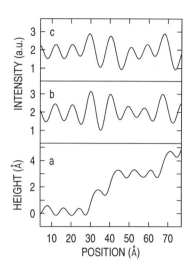

The photon intensity variations observed on flat surfaces and near steps are reproduced surprisingly well by this simple model (Fig. 3). This demonstrates that even sub-Angstrom sized topographic features can give rise to significant variations of the electromagnetic coupling of tip and sample and of the resulting photon emission intensity.

## 4. Conclusions

The above discussion demonstrates that topographic structure plays an important role in determining the photon emission intensities from atomic scale structures. In the context of near-field optics, the importance of topography effects led E. Betzig to argue that "any claims of high reolution in NSOM [near-field scanning optical microscopy] are suspect unless the features of interest can be shown to be completely uncorrelated to the topographic data" [10]. In STM-induced photon emission, the topographic information available from the STM image is sufficiently detailed to get a handle on topographical effects.

Whereas the photon emission characteristics of Au(110) discussed here can be largely understood using the topography alone, new phenomena are observed when molecules are placed between tip and sample. On $C_{60}$ on Au(110) molecular resolution in STM-induced photon emission was achieved. The emission characteristics were found to be inconsistent with purely topographic effects and, therefore, were tentatively attributed to molecular photon emission [11]. Recently, Böhringer et al. succeeded in laterally resolving the emission from single anthracene molecules on Ag(110)[12]. While the STM topographs of the anthracene molecules did not depend strongly on the direction of electron tunneling, their photon emission varied drastically. Photon emission maxima associated with the molecules were observed only when injecting electrons into the molecules whereas minima occurred under reversed bias conditions. These observations corroborate the proposal of Ref. [11] and indicate that the molecules are the primary photon source.

*Acknowledgements*

I am greatly indebted to M. Böhringer, R. Gaisch, J.K. Gimzewski, B. Reihl, R.R. Schlittler, M. Tschudy, and W.-D. Schneider for their contributions to this work. Financial support by the Swiss National Science Foundation is gratefully acknowledged.

# References

1. *Scanning Tunneling Microscopy and Related Methods*, Behm, R.J., Garcia, N., and Rohrer, H. (eds.), NATO ASI Ser. E, Vol. **184** (Kluwer, Dordrecht,1990).
2. Coombs, J.H., Gimzewski, J.K., Reihl, B., Sass, J.K., and Schlittler, R.R. (1988) J. Microscopy **152**, 325.
3. Berndt, R., Gaisch, R., Schneider, W.D., Gimzewski, J.K., Reihl, B., Schlittler, R.R., and Tschudy, M.(1995) Phys. Rev. Lett. **74**, 102–105.
4. Gaisch, R., Gimzewski, J.K., Reihl, B., Schlittler, R.R., Tschudy, M., and Schneider, W.D. (1992) Ultramicroscopy **42-44**, 1621–1626.
5. Johansson, P., Monreal, R., and Apell, P. (1990) Phys. Rev. B **42**, 9210-9213.
6. Berndt, R., Gimzewski, J.K., and Johansson, P. (1991) Phys. Rev. Lett. **67**, 3796–3799, and (1993) Phys. Rev. Lett. **71**, 3493–3497.
7. Pohl, D.W. and Courjon, D. (eds.) (1993) *Near Field Optics*, NATO ASI Ser. E, Vol. **241** (Kluwer, Dordrecht).
8. Johansson, P., Monreal, R., and Apell, P. (1993) in Ref. [7], pp. 341–352.
9. Lang, N.D. (1985) Phys. Rev. Lett. **55**, 230–233.
10. Betzig, E. in Ref. [7], pp. 7–15.
11. Berndt, R., Gaisch, R., Gimzewski, J.K., Reihl, B., Schlittler, R.R., and Schneider, W.D. (1993) Science **262**, 1425–1427.
12. Böhringer, M., Berndt, R., and Schneider, W.D. unpublished.

# WRITING OF NANOLINES ON A FERROELECTRIC SURFACE WITH A SCANNING NEAR-FIELD OPTICAL MICROSCOPE

J. MASSANELL[a], N. GARCÍA[a], A. CORREIA[a], A. ZLATKIN[a,b] and M. SHARONOV[a,b]

[a]Lab. Física de Sistemas Pequeños, CSIC-UAM, Universidad Autónoma de Madrid, C-IX, 28049 Madrid, Spain.
[b]Institute of Crystallography, Russian Academy of Sciences, 117333 Moscow, Russia.

J. PRZESLAWSKI
Departamento de Física Aplicada, Universidad Autónoma de Madrid, C-IV, 28049 Madrid, Spain.

ABSTRACT. We use a scanning near-field optical microscope in illumination mode with a metal coated optical fiber tip to write, at ambient conditions, nanolines on TGS ferroelectric surfaces. Our experiments show that, in spite of the cut-off, the outcoming light intensity is sufficient to modify locally the TGS surface. We present images that show that, with this method, linewidths as small as 60 nm can be achieved, being stable over several days.

## 1. Introduction

In the fields of biology, semiconductor technology and material science, optical microscopy remains the most important non destructive measurement technique. The resolving power of conventional far-field optics is restricted to the diffraction limit, resulting in a spatial resolution of about 0.3-0.2 μm (λ/2) [1]. This limitation is in marked contrast to the increasing interest in preparation, manipulation and characterisation of nanoscale structures with typical dimensions below 100 nm. Recently, various techniques of scanning near-field optical microscopy (SNOM) have succeeded in expanding this resolution far below the Abbe limit. A lateral resolution well beyond the diffraction limit has been claimed by several authors [2,3], however, general experience shows that 80-100 nm is a reasonable number for a typical resolution in illumination mode SNOM experiments, using metal coated fiber tips.

181

M. Nieto-Vesperinas and N. García (eds.), Optics at the Nanometer Scale 181–190.

The most important part of a SNOM apparatus is a tip that is placed in close proximity (about 10 nm) to the specimen of interest. The scanning procedure involves the movement of the specimen under the tip, using piezo translation elements. This tip can be used not only as a detector but also as a illuminating light source [4]. In this last case, the tip is used as a nanosource to illuminate the sample, allowing both materials characterisation and modification [5].

In semiconductor technology typical widths of lines and circuitry structures are in the micronscale. However larger integration of electronic devices will require nanoscale structures below the 100 nm range. SNOM can be an applicable technique to reach this goal due to the small illuminated aerea that can be achieved using metal coated fiber tips, where a small hole in the metallic cover is responsible for the dimensions of the illuminated area. Recent experiments [6] have shown that lines of less than 100 nm lateral resolution can be written in a standard multi-purpose photo resist (Hoechst Novolack AZ 6612).

In the present paper, our aim has been to write lines on a TGS ferroelectric surface in the nanoscale range (60 nm) using the SNOM fiber tip as a nanoscopic light source and using the capability of the microscope to obtain shear-force topographical images [7] to detect these lines.

## 2.    Experiment

The experiments were performed at ambient temperature using a Topometrix$^{®}$ Aurora$^{TM}$ SNOM in illumination mode. The sample is placed on top of a xyz-piezo translation stage, illuminated via a tapered fiber tip and held at constant distance by means of a shear force measurement. After illumination, the shear force topographical image was taken in the same region in order to observe the material modifications. The fiber tip is drawn with a commercial micropipette puller in order to obtain a tip radius of the order of 100 nm. Then, it was subsequently coated with about 70 nm Al. The beam of an $Ar^{+}$ ion laser ($\lambda = 488$ nm) is coupled into a single mode optical fiber.

The nanowriting attempt has been performed on cleaved TGS [triglycine sulphate, $(NH_2CH_2COOH)_3 - H_2SO_4$] sample. The crystal was grown from water solution at ambient temperature which corresponds to the ferroelectric phase ($T_c = 49.7$ C). The microscopic structure of the cleaved surface is quite similar to that presented in Refs. 8 and 9: a sequence of microscopic terraces that can be also seen in figures 1 and 2.

3.    **Results**

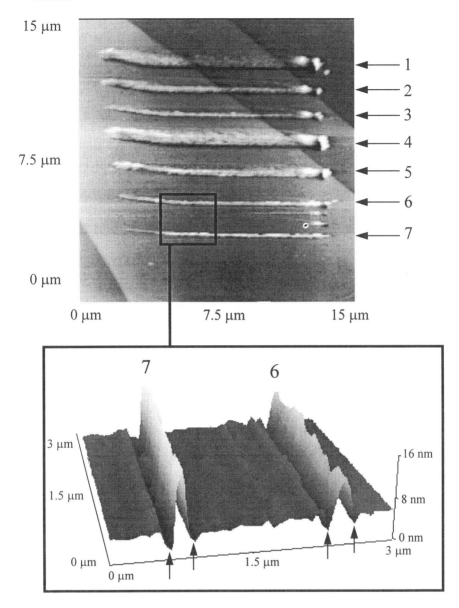

*Fig.1.* Shear force topographical image of nanolines in a TGS surface. Notice the 3 steps clearly seen in the micrograph. The mean step height corresponds to the height of two unit cells (2×1.26 nm). The zoom shows part of the thinnest lines in 3D-presentation. Notice the valleys beside lines.

Figure 1 shows typical lines that can be written as described above. These lines are observed in the shear force topographical images and could not be found in the optical images [10]. The lines are labelled in the figure from 1 to 7. Lines as small as 120 nm width can be obtained with a single illumination scan (see figures 1 and 2). Thicker lines can be drawn by several adjacent illuminating scans.

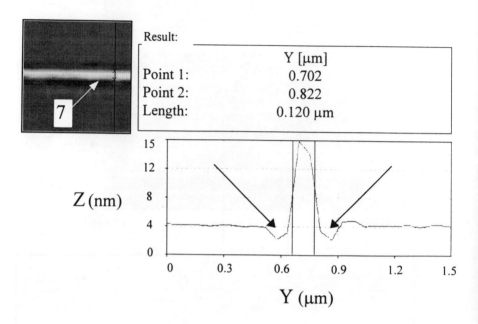

Fig. 2. Cross-section view of line 7 showing a linewidth of 120 nm. Notice the valleys beside the line, marked by the arrows.

The line thickness measured is enlarged with respect to the real thickness due to the effect of the tip geometry on the image. This enlargement can be estimated from known topographic features like steps. We can then correct the measured linewidths. In table 1 we present the directly read and also the deconvolved line widths. We obtain that the smallest line has a width of about 60 nm, which is also in good agreement with the 80 nm aperture diameter. Under these conditions the light power through the tip aperture is typically in the range of 1 nW to 100 nW.

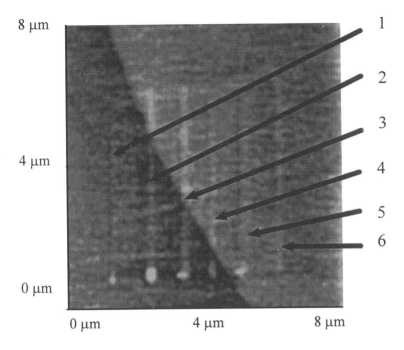

8 µm

4 µm

0 µm

0 µm          4 µm          8 µm

1

2

3

4

5

6

*Fig. 3.* Shear force topographical image of vertical nanolines in a TGS surface. Notice the double step at x ≈ 3.2 µm from the upper left to the bottom right side.

In figure 3 the same lines are drawn but in the vertical direction. Their characteristics are quite similar to those of figure 1 and are indicated in table 1, although now the structures are in a larger noise background. In addition, by increasing the laser intensity by a factor of about 5, we obtain thicker structures in the micrometer range, and now they can be seen both in the shear force topography (fig. 4a) and in the optical image (fig 4b). All these lines (figs. 1-4) were stable during three days without noticeable modifications.

Table 1: Directly measured (averaged) and deconvolved linewidth values of the nanolines shown in figure one and two. The numbers in parenthesis indicate the number of adjacent illuminating scans.

| | Linewidth [nm] measured from shear force map figure 1 | Deconvolved linewidth [nm] | Linewidth [nm] measured from shear force map figure 2 | Deconvolved linewidth [nm] |
|---|---|---|---|---|
| Line 1 | ~ 500 (10 scans) | ~ 440 | ~ 125 (1 scan) | ~ 65 |
| Line 2 | ~ 255 (5 scans) | ~ 195 | ~ 200 (3 scans) | ~ 140 |
| Line 3 | ~ 140 (1 scan) | ~ 80 | ~ 220 (3 scans) | ~ 160 |
| Line 4 | ~ 620 (10 scans) | ~ 560 | ~ 175 (1 scan) | ~ 115 |
| Line 5 | ~ 350 (5 scans) | ~ 290 | ~ 250 (3 scans) | ~ 190 |
| Line 6 | ~ 120 (1 scan) | ~ 60 | ~ 130 (1 scan) | ~ 70 |
| Line 7 | ~ 120 (1 scan) | ~ 60 | | |

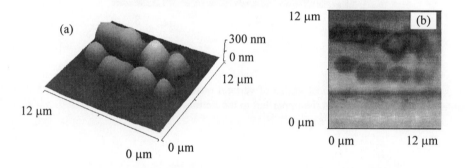

*Fig. 4.* Thick lines in the micronscale. (a) shear force topographical image in 3D-presentation; (b) optical image measured in reflection. The contrast is about 5 %.

## 4. Discussion

The question that arises is: what mechanism is responsible for the appearance of the lines? In our opinion, several mechanisms may be responsible for this writing process:
1) Charge removal. Given the ferroelectric nature of the material there is an electric field at the surface of the order of $10^6$ to $10^7$ V/m. This field [11] is not strong enough to eject

electrons. However, surface electrons could be excited by illumination. In these conditions, the field emission barrier of the photo excited electrons would decrease, and emission would be more probable. In addition, lateral motion of electrons, that have been photo excited to higher conducting levels at the surface, could occur. This charge inhomogenety created by the light would induce image charges in the metal coated tip. The coulomb interaction would lead to local force gradients, affecting the shear force feedback signal.

2) Photo chemistry of the ferroelectric TGS surface due to its organic characteristics. The illumination may induce bond breaking and surface photo chemistry leading to lines of new material. A smaller density of the modified material would lead to humps in the surface.

3) Although intensity of the light is not enough to heat the surface, it is possible that the tip heats up due to energy dissipation in the cut off region. A hot tip could increase the surface temperature locally leading to TGS decomposition.

While the first mechanism may be responsible for the smaller lines, we notice that the lines could not be observed in optical images. The second or third mechanism may be responsible for the larger structures. At present these points are under investigation.

To estimate if it is possible to measure charge effects in the feedback signal with our experimental configuration, we consider a simple model, shown in figure 5:

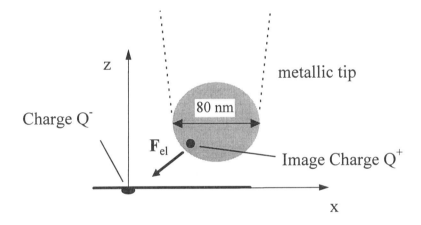

*Fig. 5.* Model that allows a rough estimation of the coulomb force gradient due to charge - image charge interaction.

A point charge, located at x=0 z=0, creates an image charge in the metallic tip, modelled as a metallic sphere with a radius of 80 nm. The coulomb force gradient in the x-direction, $(dF_x/dx)$, due to the interaction between a $1.6 \ 10^{-16}$ C charge and its image, in the metallic sphere placed at a height of 20 nm, is displayed in figure 6 as a function of the tip -

charge horizontal distance, x. Figure 7 shows the dependence of this force gradient on the tip-sample vertical distance, z.

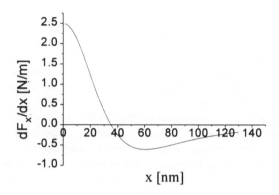

*Fig.6.* Force gradient as a function of the horizontal displacement of the tip, calculated in the case of a charge of $1.6 \ 10^{-16}$ C and a tip moving at 20 nm height.

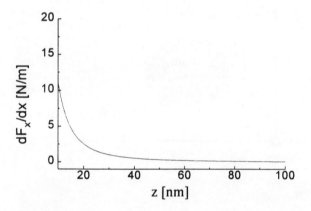

*Fig. 7.* Force gradient as a function of tip - sample vertical distance calculated with our model for a $1.6 \ 10^{-16}$ C charge. The charge is located right under the tip. Notice the decay length between 10 and 20 nm.

The force gradient in the x-direction changes the cantilever resonant frequency, affecting the feedback signal. The force constants of the fiber tips used in our experiments are in the range of some hundreds of Newtons per meter. The Q-factor is typically about a hundred. Accordingly, the dynamic range to measure force gradients with this configuration is in the range of some Newtons per meter.

The model produces the following results:

1) The order of magnitude of the calculated force gradients is about 10-40% of the dynamic range of the considered non-contact measurement, i.e., they should be visible in the shear-force topography.

2) The force gradient changes sign at a tip horizontal x-displacement of a few tens of nanometers. This result could explain the valleys beside the lines observed in our experimental data. (Figs. 1 and 2)

3) The decay length of the calculated force gradients is in the same range than the typical decay length of the shear force signal, making non trivial a differentiation between electric and topographical contrast.

4) A charge located at one point leads to a linewidth of about some tens of nanometers.

These results point out in the direction of charge inhomogeneties as the cause of the observed contrast.

## 5.    Summary

A scanning near-field optical microscope in illumination mode has been used to write nanolines down to 60 nm line width on TGS surfaces. It is interesting to note that although there is a cut-off in the emitted light due to the metal cover that has an aperture smaller than $\lambda/2$, there is sufficient light intensity to write and modify the surface locally below the 100 nm range. Several mechanisms that could be responsible for this effect have been proposed.

Acknowledgements:

The authors would like to thank the workgroups of J.Mlynek (University of Constance) and O.Marti (University of Ulm) for providing us with metal covered optical fiber tips. Helpful discussions with A.P. Levanyuk and J.L. Costa-Krämer are gratefully acknowledged. This work has been supported by the DGICYT project PB94-0151 and the NATO Linkage Grant: NANO.LG.940 557.

190

References:

[1]    E. Abbe, Schulzes Archiv f. mikr. Anat 9, 413 (1873).
[2]    D.W. Pohl, W. Denk, M. Lanz, Appl. Phys. Lett. **44**, 651 (1984)
[3]    E. Betzig, J.K. Trautman, T.D. Harris, J.S. Weiner and R.L. Kostelak, Science **251**, 1468 (1991) and Science **257**, 189 (1992).
[4]    See *Near-Field Optics*, ed. by D.W. Pohl, D. Courjon, NATO ASI Ser. E, Vol. 242 (Kluwer, Dordrecht 1993)(p.7)
[5]    J. Massanell, N. García, A. Zlatkin, Optics Letters (in press)
[6]    G. Krausch, S. Wegscheider, A. Kirsch, H. Bielefeldt, J.C. Meiners, J. Mlynek, Optic Comunications (1995) (in press)
[7]    R. Toledo-Crow, P.C. Yang, Y. Chen and M. Vaez-Iravani, Appl. Phys. Lett. **60** (24), 2957 (1992).
[8]    N. Nakatani, J. Phys. Soc. Jpn **39**, 741 (1975).
[9]    N. García, A.P. Levanyuk, J. Massanell, J. Przeslawski, A. Zlatkin and J.L. Costa-Krämer, report made at the 8th european meeting on ferroelectrics, Niemegen, July 1995 and to be published in Ferroelectrics.
[10]   This may be due to low signal to noise ratio or that the small lines are not topographical structure but possibly written charge distribution inhomogenities.
[11]   P. Wurfel and I.P. Batra, Phys. Rev. B **8**, 5126 (1973).

# NEAR FIELD OPTICS
# WITH HIGH-Q WHISPERING-GALLERY MODES

N. DUBREUIL, J. C. KNIGHT, J. HARE,
V. LEFEVRE-SEGUIN, J. M. RAIMOND and S. HAROCHE
*Laboratoire Kastler-Brossel*[†]
*Département de Physique, Ecole Normale Supérieure*
*24 Rue Lhomond 75005 Paris France*

High-Q whispering-gallery modes (WGM) are resonant electromagnetic modes which can be observed in dielectric microspheres with diameters larger than about ten times the wavelength [1]. They correspond to high angular momentum states of the electromagnetic field in which light propagates close to the sphere's surface by repeated total internal reflection. The confinement of the light field in a small volume implies very high values of the electric field per photon. As a result, microdroplets have been shown to exhibit various non-linear optical effects induced at low power excitation. Cavity-enhanced stimulated emission [2], as well as laser action [3], has also been observed in dye-doped droplets. More recently, semiconductor microcavities have been investigated and microdisk lasers based on WGMs have been demonstrated [4]. Another remarkable feature can be achieved with silica microspheres. In this low-loss material, photon storage times in the range of one microsecond have been observed, which correspond to quality factors Q greater than $10^9$ [5, 6, 7, 8]. Such high Q values combined with small mode volumes are rather promising for the pursuit of Cavity Quantum Electrodynamics experiments in the strong coupling regime where a few atoms or ions are tightly coupled to a cavity mode which contains a small number of photons, or is even in the vacuum state [9]. These two properties have also allowed several groups to observe laser action in doped microspheres [10, 11] and we are currently investigating high-Q neodymium-doped microspheres showing very low-thresholds.

[†]Laboratoire de l'Ecole Normale Supérieure et de l'Université Pierre et Marie Curie, Associé au CNRS (URA 18)

*M. Nieto-Vesperinas and N. García (eds.), Optics at the Nanometer Scale 191–203.*
© *1996 IBM. Printed in the Netherlands.*

For all practical experiments, it is desirable to know precisely the field distribution associated with the mode being excited. When microlaser performance is investigated, for instance, the spatial overlap between the pump mode and the lasing mode influences the threshold power and the mode competition. Using a fiber tip as a near-field probe, we have recently been able to map directly the different field patterns associated with various modes [12, 13]. The principle of this measurement and the experimental details will be described in the next sections. Further applications of this method will be discussed in the conclusion.

## 1. Whispering-gallery mode patterns

In these modes, first conceived by Lord Rayleigh [14] and analyzed later by G. Mie [15], light travels at grazing incidence in nearly great circles, being trapped at the surface by the discontinuity of the refractive index $N$. The complete description of these modes [1] is derived from the solution of the Maxwell's equations with the appropriate boundary conditions. Resonances occur at discrete values of the size parameter $x = ka = 2\pi a/\lambda$, where $k$ is the wavenumber and $\lambda$ the wavelength in vacuum. Each WGM is also described by its polarization (TE or TM) and a set of three mode numbers: a radial number (or order) $n$ which gives the number of maxima in the radial dependence of the mode, an angular mode number $l$ (roughly giving, for low $n$ modes, the number of wavelengths in a round trip inside the spherical cavity) and an azimuthal mode number $m$ which can take the values $m = -l,...0,...+l$. The mode numbers $l$ and $m$ describe the angular momentum of the mode. Opposite values of $m$ correspond to waves running in opposite directions around the sphere's center. In a perfect sphere, modes differing only in $|m|$ would be degenerate, as great circles inclined at different angles are the same. However, microdroplets, or microresonators like ours which are made by heat-fusing the tip of a silica fiber, will always show some departure from the spherical shape. Instead of great circles, the classical orbits will rather be (to the lowest order) ellipses slowly precessing around the axis of symmetry of the resonator. This axis approximately corresponds in our case to the stem of the fiber to which it is attached. This changes the optical path and breaks the degeneracy between resonances with different $|m|$ values [16]. For small ellipticities, the angular variation of the intensity inside the microsphere $I_{lm}(\theta, \phi)$ at a size parameter $x_{nlm}$ corresponding to a $n, l, m$ resonance of either polarization can be expressed using spherical harmonics. By using approximations for large $l$ and for $|m| \simeq l$ we have shown that it can be approximated as

$$I_{lm}(\theta, \phi) \propto |H_{l-|m|}(l^{1/2} \cos \theta) \sin^{|m|}(\theta) \exp(im\phi)|^2 \qquad (1)$$

where $H_{l-|m|}$ represents a Hermite polynomial, $\theta$ is the polar angle and $\phi$ is the angle in the equatorial plane. From this equation a $n, l, m$ mode has $l - |m| + 1$ major lobes in its polar intensity dependence, centred on $\theta = \pi/2$ and extending out to $\theta = \pi/2 \pm \arccos(m/l)$, $\theta = 0$ being defined by the symmetry axis of the deformed spheroid. When the two opposite directions of propagation corresponding to $\pm m$ are equally excited, the field intensity exhibits $2|m|$ lobes along the sphere's equator. In order to minimize the mode volume while maintaining the very high Q achievable with fused-silica microspheres it is desirable to excite modes with $|m|$ close to $l$: such modes have only a few lobes and are closely confined to the equatorial region of the sphere ($\theta = \pi/2$). Although degeneracy breaking of WGM's has previously been observed in stimulated Raman scattering from liquid droplets flowing in air [17, 18, 19], and has been deduced from photographs of the near-field patterns formed on the surface of a prism used to couple to WGM's in a fused-silica sphere [20], in neither case has the specific form of the degeneracy-broken modal field patterns been verified. In a first set of experiments described in section 2, we mapped the field distribution along a meridian. We then excited a stationary equatorial pattern as a superposition of two counter-propagating modes in order to observe the field distribution along the equator.

## 2. Mapping the near-field distribution of WGMs: experimental details

We have investigated the angular dependence of different high-Q WGM's excited in silica microspheres in the size range $D=2a=180\text{-}450\mu\text{m}$ by using a near-field optical probe. The microsphere is formed at the end of a silica wire or fiber with a $CO_2$ laser or a microtorch. Fig. 1 shows a SEM picture of such a microresonator. The microsphere modes are probed with

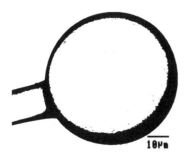

*Figure 1.* SEM photography, at low resolution, of a silica microsphere at the end of the preform fiber. Its diameter is 70 $\mu$m. At higher resolution, no surface defect is observed on a 30 nm scale.

a submicron-sized tip formed at the end of a length of monomode optical fiber. When such a tip is placed in an evanescent field formed by total internal reflection, the reflection is frustrated and light is coupled into the optical fiber. This effect has been used in the development of the near-field optical microscope [21] in which the tip is placed in an evanescent light field near a surface and then scanned to provide a surface map of morphological or refractive-index features. It was recently shown that a near-field optical probe could be used to map the electromagnetic field inside a Fabry-Perot resonator [22] by folding the resonator on a total internal reflection prism and mapping the evanescent field on the prism surface. In a similar spirit, by moving such a probe in the evanescent field at the surface of a resonant microsphere we have mapped the angular dependence of the mode patterns for different resonance modes.

Efficient excitation of WGM's in microspheres is achieved by frustrated total internal reflection, either with a high-index prism [5, 6] or with a polished monomode fiber [23]. The experiments described here were performed with this type of fiber-coupler based on a monomode optical fiber mounted on a substrate and side-polished so as to expose the field of the propagating mode. Details of the experiment are shown schematically in Fig. 2. The polished input fiber was oriented so as to couple light from the fiber to the equatorial region of the sphere. The gap between the sphere and the input fiber was controlled using a microcontrol and a PZT positioner. Light from a scanable grating-stabilized diode laser operating at around 786nm was coupled into the fiber, with its polarization set so as to excite transverse electric (TE) or transverse magnetic (TM) resonances. Although the fiber used to make the coupler is not polarization-preserving, the polarization is largely maintained over the short length of fiber used in the experiment.

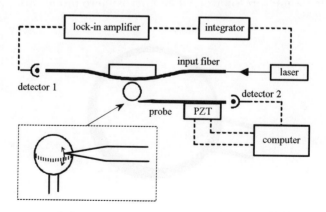

*Figure 2.* Schematic diagram of the experimental set-up; the inset shows the sphere-probe region for the polar mapping experiment.

When the diode laser was scanned past a microsphere resonance light was coupled out of the fiber and into the microsphere. By using a lock-in amplifier to detect the absorption dip in the light leaving the input fiber (signal from detector 1 in Fig. 2) the laser could be electronically locked to the resonance.

Our probes were fabricated by placing the end of a length of monomode optical fiber in a hydrofluoric acid solution (40%) for a period of 80 minutes. By using this method we were easily able to produce tips which had a radius of less than a micron, which gave a sufficient resolution for the present application. The method of mapping evanescent fields was tested by placing the probe directly on the polished surface of the monomode input fiber and moving the probe under piezoelectric control across the exposed evanescent field of the propagating fiber mode. By placing the probe on the surface with an angle of about 20 degrees between the axis of the polished fiber and that of the probe fiber we were able to record a position-dependent intensity signal from the exit of the probe fiber which was in excellent agreement with the calculated spatial dependence of the guided fiber mode.

The fiber tip to be used as a probe of a whispering-gallery mode was placed opposite the input fiber in the plane of the sphere's equator. It was mounted on two levered PZT stacks, which allowed it to be brought into contact with the sphere and then moved on the surface. This was done by amplifying appropriate signals from a computer to drive the two probe PZT's. At the same time the light from the exit port of the probe fiber was detected using a photodiode (detector 2) and the signal (typically of the order of $0.1\mu$W) recorded by the computer. Characterizing the spatial distribution of a WGM was done in two steps. The probe fiber was first set in the equatorial plane as shown in Fig. 2, lying nearly along the equator. It was then moved along a meridian to sample the WGM field out of the equatorial plane and identify $l - |m|$.

In a second step, the setup was slightly modified to excite a standing-wave in the mode (see Fig. 3). A dielectric mirror was butt-coupled to the end of the input fiber (approximately 25 cm after the fiber coupler) with a drop of index-matching liquid, resulting in almost total reflection of the light in the fiber. Because we couple only a few percent of the light in the fiber into the sphere this results in two almost equal counterpropagating fields in the input fiber, giving a standing wave in the microsphere resonance with $2|m|$ intensity maxima in its angular dependence along the sphere's equator. To get enough lateral resolution, the fiber tip was set nearly perpendicular to the equatorial plane (see Fig. 3). Although this is not favourable in principle for matching the wave-vectors of the WGM to the guided mode in the tip fiber, we observed enough transmission, most probably because of the roughness of the tip end. The tip holder was designed to be one of the arms of an interferometer to provide a calibration of the displacement of the tip. The interferometer signal and the signal from the probe output were detected using photodiodes and recorded simultaneously by the computer driving the piezoelectric transducers.

## 3. WGM field mapping along a meridian

Scanning the field distribution along a meridian allowed us to observe different mode patterns. Some results obtained on a sphere of size $2a=286\mu m$ are shown in Fig. 4 (thick lines). These are selected from a series of modes observed to have from 1 to more than 12 polar lobes: those plotted in Fig. 4 are the observed angular patterns of modes displaying 1, 2, 4, 6 and 8 polar lobes, corresponding to $l-|m|=0$, 1, 3, 5 and 7. Also plotted are fits to the data of equation (1) (thin lines), where we have varied $l$ as a free parameter (keeping $l-|m|$ fixed). The data are well fitted by a function of the form of equation (1). While the number of lobes in the angular dependence of each mode can be easily and unambiguously identified from plots such as those shown in Fig. 4, the relative heights of the different peaks are to a certain extent dependent on the probe's accurately following the curved sphere surface, and on the effect of the probe on the mode. The data presented in Fig. 4 were recorded in the overcoupled case (sphere in contact with the input coupler) but we have successfully applied the method in the undercoupled regime as well. In this case the observed resonance linewidth can be of the order of 1 MHz (limited by the linewidth of our diode laser), corresponding to a $Q > 10^8$. Because the stem on which the sphere is formed has a diameter of about $60\mu m$, it is possible to move the $1\mu m$ probe about on the surface of the sphere without significantly affecting the relative posi-

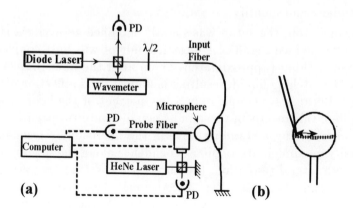

*Figure 3.* (a): Schematics of the equatorial mapping experimental set-up; (b): details of the sphere-probe region.

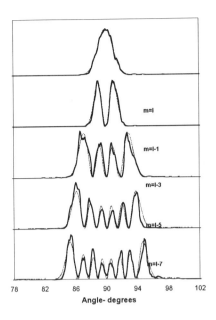

78    82    86    90    94    98    102
**Angle- degrees**

*Figure 4.*    Polar field distribution observed from different resonant TE modes in a fused-silica microsphere of size $2a = 286$ $\mu m$ (thick lines) and the fitted patterns from Eq. (1) (thin lines). The actual intensities have been independently scaled and offset for the purposes of the figure.

tions of the sphere and the input fiber. Any decrease in the sphere/fiber gap would be observable as a broadening and a shift of the narrow resonances. The presence of the fiber tip near an antinode in the mode volume itself causes a slight shift and broadening of the resonance, but the resonance width remains less than 10MHz.

By scanning the wavelength of our diode laser we were able to observe different resonances and we have measured the number of lobes in the angular pattern of each of them. The resonant wavelengths (measured with a wavemeter) are plotted in Fig. 5 as a function of the observed number of lobes. These are seen to form a number of almost straight lines with very similar slopes. The spacing between the modes on one of these lines is approximately 0.015nm, and corresponds to the splitting of the degenerate sphere modes by the asphericity in our microcavity. In contrast, in a sphere of this size, the spacing between TE modes with the same $n, l - |m|$ values and with $l$ differing by one, defined as $\Delta\lambda_{FSR}$, is $\simeq 0.529$nm [24]. Modes separated by this amount are plotted using the same symbols in Fig. 5. Our excitation field from the input fiber has a very small angular range which together with the similarity of the indices of the sphere and of the input fiber is expected to result in our exciting only the modes with the lowest

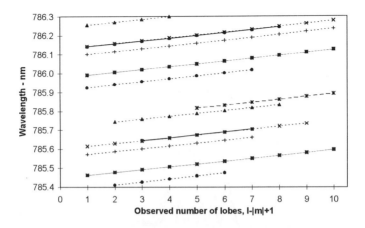

*Figure 5.* Observed TE resonance wavelengths plotted as a function of the number of lobes observed in the angular dependence for each mode. The lines join the points as an aid to the eye. Those lines which are not complete correspond to resonances which are not strongly excited and which could not all be unambiguously mapped. Lines plotted with the same symbols show modes with the same $n$-value but with $l$ differing by one. The precision of the measurement of the wavelength of a particular mode is approximately 0.0005 nm.

radial orders, which correspond to $l \simeq n_s x$. Accordingly, we observed that within an FSR, only two modes with $l - |m| = 0$ were strongly excited. This was also true for TM modes. The value of $l$ obtained from the fits to the data shown in Fig.4 (which were recorded from a single line in Fig. 5, i.e. have the same $n, l$ values) is around 1650, consistent with the excitation of low-$n$ modes, and vary by less than 5%. For small deviations from sphericity, for large $l$, and for $|m| \simeq l$, the slope of the line in Fig. 5 is $\Delta\lambda/\Delta l - |m| = e\lambda/l$, where $e = (a_p - a_e)/a$ is the ellipticity of the microparticle in terms of the polar, equatorial and volume-average radii [16]. Assuming $l = 1650$ this gives a figure of $e = 0.03$ for the sphere. The positive sign of $e$ corresponds to a prolate deformation which is consistent with a deformation of the sphere during fabrication due to the presence of the stem. The different lines shown in Fig. 5 are not perfectly parallel because they correspond to the excitation of different $n, l$ values. The data shown in Fig. 5 were recorded without changing the angle of the input fiber with respect to the sphere. As a result, larger $l - |m|$ modes are increasingly weakly excited and difficult to observe.

These experiments have allowed us to measure the extension of a given mode out of the equatorial plane and determine $l - |m|$ unambiguously.

However, the angular mode number is not known well enough to determine the radial mode number. Mode identification is usually performed by scanning a large number of FSRs and fitting the observed resonance positions with the theory [24, 25]. This remains difficult here, partly because high-order shape effects could affect the mode spectrum. We show in the next section how to identify the radial mode number from the equatorial mode pattern.

## 4. Equatorial WGM standing-wave pattern

In a second series of measurements, we mapped the field variation along the equator. As explained above, a resonant mode with $m = l$ was excited in both directions so as to form a standing wave in the equatorial plane with $2m$ intensity maxima along the sphere perimeter. The fiber tip, set nearly perpendicular to the equatorial plane (see Fig. 3), was used to probe this periodic pattern. The fiber tip intensity and the interferometer signals were recorded simultaneously as a function of the PZT voltage driving the tip displacement.

The experimental results to be presented here were recorded from a sphere made of pure fused silica (refractive index N=1.45356 at $\lambda$=788nm). Measurement of the sphere by optical microscopy gives the equatorial diameter as $2a \simeq 280\mu$m, corresponding to a value of the size parameter $x = 2\pi a/\lambda$ of $x \simeq 1120$. Modes with only a single lobe in their polar dependence, i.e. fundamental WGM's with $l = |m|$ were selected and from here on we restrict our discussion to such modes. The observed separation between modes with the same value of the radial mode number and with a difference of unity in the angular mode number (equivalent to the free spectral range or FSR of the microresonator) is $\Delta\lambda_{FSR} = 0.491$nm. By tuning the laser over one FSR we find that there are two WGMs of each polarization (corresponding to two different radial mode numbers) which are very strongly excited by the fiber coupler, and several which are somewhat less strongly excited. The observed spacing between modes with the same mode numbers and differing only in polarization is $\Delta\lambda_{TE-TM} = 0.351$nm. These values for the observed $\Delta\lambda_{FSR}$ and $\Delta\lambda_{TE-TM}$ can be used to improve the determination of $a$ which is found to be $a = 140 \pm 0.5\mu$m.

A portion of a record is shown in Fig. 6 for one of the most strongly excited TE mode. A typical set of data covered approximately $30\mu$m and included more than 100 antinodes of the standing wave. We verified that the plane of translation of the fiber tip corresponds well (to within 2 degrees) with the plane of the whispering-gallery mode. Fig. 6(b) shows the fast Fourier transforms of the two signals. The spectra are highly monochromatic, allowing an accurate measurement of the period of the standing wave

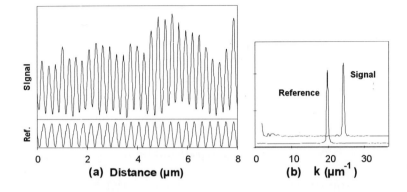

*Figure 6.* Equatorial mapping : (a) the fiber tip intensity signal is recorded as a function of the tip position along the equator, calibrated with the interferometer reference signal. Only a small portion of a typical record is shown; (b) Fast-Fourier transforms of the signal and reference used to determine their spatial frequencies.

in the microsphere mode by comparison with the known wavelength of the HeNe laser. This directly gives the ratio $|m|/a = l/a$. Since the resonant wavelength is measured independently with a wave-meter, we get an experimental determination of the ratio $l/x_{nlm}$. The final step to identify the radial mode number $n$ is to compare the experimental value of this ratio with numerical computations for several values of $n$. This comparison is shown in Fig. 7 for the two most strongly excited TE modes. Our data are in good agreement with $n = 1$ and $n = 2$. The determination of $n$ combined with the knowlegde of $m$, $l$ and $x$ within 1% allows us to estimate the mode volume with the same precision. This is important to determine the field amplitude per photon in the mode, which is crucial to estimate the coupling strength of atoms or ions to a WG mode.

## 5. Conclusion

Mapping WGM modes in the near-field have allowed us to determine with a good accuracy the three quantum numbers of a given resonance. Our work also shows that the use of a fiber coupler to excite WGM's in fused-silica microspheres results in the strongest excitation of those modes with the lowest radial mode numbers and the lowest mode volumes. This makes the fiber coupler an ideal compact excitation source for Cavity Quantum

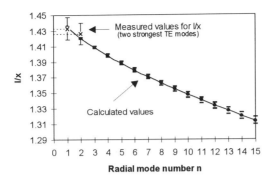

*Figure 7.* Theoretical values of $l/x$ are shown as a function of $\nu$. The bars show the range of calculated values for all resonances compatible with the experimental uncertainty of 5% on $x$. The experimental values of $l/x$ for the two strongest TE modes with $l = |m|$ are indicated by crosses on the vertical axis. The horizontal lines are an aid to the eye. The experimental uncertainty on $l/x$ of about 5% is enough to produce an overlap with two possible $\nu$ values for each mode. As expected for low radial numbers, the experimental points lie close to the value of $n_s$ ($n_s = 1.4545$ for pure silica at 790 nm).

Electrodynamics experiments where strong coupling betweem atoms or ions and the electromagnetic field is needed. The use of fibercouplers and fibertips will also facilitate some of these projects which involve a liquid helium environment to obtain the minimum transition linewidth. We are currently investigating the properties of microlasers based on the whispering-gallery modes of Nd-doped silica microspheres. Using a fibertip to ascertain the polar and equatorial WGM field distribution and deduce its radial confinement will certainly be a precious tool in these experiments where laser action strongly depends on the spatial overlap of the pump and lasing mode. Furthermore we believe that the results reported here demonstrate the possibility of placing small objects at a desired point in the mode volume, which ability will be useful to further study the interaction of atoms or ions with these microscopic cavities.

**Acknowledgements:** The authors gratefully acknowledge the support of a France Telecom CNET/CNRS contract. We thank Michel Gross for help and advice with the use of the wavemeter. Laboratoire Kastler Brossel is a laboratory of the Université Pierre et Marie Curie and Ecole Normale Supérieure, associated with the Centre National de la Recherche Scientifique. N. Dubreuil is also at the Laboratoire d'Optronique associé au Centre National de la Recerche Scientifique (EP 001), Université de Rennes, Ecole Nationale Supérieure de Sciences Appliquées et de Technologie, 6 rue de Kérampont, 22305 Lannion, France.

202

# References

1. A thorough review of Mie resonances is presented in *"Optical effects associated with small particles"*, Advanced Series in Applied Physics, Vol.1, P. W. Barber and R. K. Chang, Eds. (World Scientific, Singapore, 1988).
2. Campillo A. J., Eversole J. D. and Lin H. -B., (1991) Cavity Quantum Electrodynamics enhancement of stimulated emission in microdroplets, *Phys. Rev. Lett.* **67**, 437.
3. Tzeng H. -M., Wall K. F., Long M. B. and Chang R. K., ( 1984) Laser emission from individual droplets at wavelengths corresponding to morphology-dependent resonances, *Opt. Lett.* **9**, 499.
4. Yamamoto Y. and Slusher R. E., Optical Processes in Microcavities, *Physics Today*, **46**(6), June 1993, p. 66, and references included.
5. Braginsky V. B., Gorodetsky M. L. and Ilchenko V. S., (1989) Quality-factor and nonlinear optical properties of optical whispering-gallery modes, *Phys. Lett.* **A 137**? 393.
6. Collot L., Lefèvre-Seguin V., Brune M., Raimond J. M., and Haroche S., (1993) Very high-Q resonances observed on fused silica microspheres, *Europhys. Lett.,* **23**, 327.
7. Weiss D. S., Sandoghdar V., Hare J., Lefèvre-Seguin V., Raimond J. -M. and Haroche S., (1995) Splitting of high-Q Mie modes by light backscattering in silica microspheres, *Opt. Lett.* **20**, 1835.
8. Gorodetsky M. L., Savchenkov A. A. and Ilchenko V. S., (1995) Postdeadline paper submitted to *ICONOL Laser Optics 1995, St. Petersburg.*
9. Haroche S., Cavity Quantum Electrodynamics, *"Fundamental Systems in Quantum Optics, Les Houches Summer School, Session LIII"*, J. Dalibard, J.M. Raimond, and J. Zinn-Justin, eds., (North Holland, Amsterdam, 1992) and references included.
10. Yuzhu Wang, Yongqing Li, Yashu Liu and Baolong Lu, (1991) A Nd-glass microsperical cavity laser induced by Cavity QED effects, in *SPIE Proc. Advanced Laser Concepts and Applications, Vol. 1501* p. 40.
11. Kuwata-Gonokami M., Takeda K., Yasuda H. and Ema K., (1992) Laser emission from dye-doped polystyrene microspheres, *Jpn. J. Appl. Phys.* **31**, L99.
12. Knight J. C., Dubreuil N., Sandoghdar V., Hare J., Lefèvre-Seguin V., Raimond J. -M. and Haroche S., (1995) Mapping whispering-gallery modes in microspheres with a near-field probe, *Opt. Lett.* **20**, 1515.
13. Knight J. C., Dubreuil N., Sandoghdar V., Hare J., Lefèvre-Seguin V., Raimond J. -M. and Haroche S., (1996) Characterising whispering-gallery modes in microspheres by direct observation of the optical standing wave in the near field, submitted to *Opt. Lett.*.
14. Lord Rayleigh, (1877)"The Theory of Sound", vol.2 Chap.14, (Dover Publications, N. Y. , 1965); see also Lord Rayleigh, (1881) *Phil. Mag.* **12**, 81 and (1914) *Phil. Mag.* **27**, 100.
15. Mie G., (1908) *Ann. Phys.* **25**, 377
16. Lai H. M., Lam C. C., Leung P. T.and Young K., (1991) Effect of pertubations on the widths of narrow morphology-dependent resonances in Mie scattering, *J. Opt. Soc. Am.* **B8** 1962.
17. Chen G., Chang R. K.,. Hill S. C, and Barber P. W., (1991) Frequency splitting of degenerate spherical cavity modes: stimulated Raman scattering spectrum of deformed droplets, *Opt. Lett.* **16**, 1269.
18. Swindal J. C., Leach D. H., Chang R. K., and Young K., (1993) Precession of morphology-dependent resonances in non-spherical droplets, *Opt. Lett.* **18**, 191.
19. Chen G., Mazumder M. M., Chemla Y. R., Serpengüzel A., Chang R. K., and Hill S. C., Wavelength variation of laser emission along the entire rim of slightly deformed microdroplets, *Opt. Lett.* **18** (1993) 1993.
20. Gorodetsky M. L. and Ilchenko V. S., (1994) High-Q whispering-gallery microres-

onators: precession approach for spherical mode analysis and emission patterns with prism couplers, *Opt. Commun.* **113** 133.

21.  Heinzelmann H. and Pohl D. W., (1994) Scanning near-field optical microscopy, *Appl. Phys. A* **59** 89.

22.  Courjon D., Bainier C. and Baida F., (1994) Seeing inside a Fabry-Perot resonator by means of a scanning tunneling optical microscope, *Opt. Commun.* **110**, 7.

23.  Dubreuil N., Knight J. C., Leventhal D. K., Sandoghdar V., Hare J., and Lefèvre V., (1995) Eroded monomode optical fiber for whispering-gallery mode excitation in fused-silica microspheres, *Opt. Lett.* **20** , 813.

24.  Schiller S. and Byer R. L., (1991) High-resolution spectroscopy of whispering-gallery modes in large dielectric spheres, *Opt. Lett.* **16**, 1138.

25.  Eversole J. D., Lin H. -B., A. L. Huston and Campillo A. J., (1990) Spherical cavity-mode assignments of optical resonances in microdroplets using elastic scattering, *J. Opt. Soc. Am.* **A7**, 2159.

# FLUORESCENCE MICROSCOPY AND SPECTROSCOPY BY SCANNING NEAR-FIELD OPTICAL / ATOMIC FORCE MICROSCOPE (SNOM-AFM)

MASAMICHI FUJIHIRA
*Department of Biomolecular Engineering,*
*Tokyo Institute of Technology,*
*4259 Nagatsuta, Midori-ku, Yokohama 226, Japan*

ABSTRACT. We developed a new scanning near-field optical microscope (scanning near-field optical / atomic-force microscope; SNOM-AFM). In SNOM-AFM, we took advantage of non-contact atomic force microscopy (AFM) to control the tip-sample distance for scanning near-field optical microscopy (SNOM) without mechanical damages of the tip and sample surfaces. By the precise control of the distance of the optical fiber tip within 100 nm from the sample surface, we succeeded in observing fluorescence micrographs and also fluorescence spectra for localized micro-areas of various samples. These results observed previously will be summarized in this review.

## 1. Introduction

Various types of microscopes, such as the fluorescence microscope [1-3], Brewster angle microscope [4,5], transmission electron microscope (TEM) [6-8], scanning tunneling microscope (STM) [9], and AFM [10,11], have been used to study the morphology of monolayer films. A problem of the former two optical microscopes is limited spatial resolution on the order of wavelengths, while a shortfall of the latter three microscopes is lack of chemical recognition ability. To cope with these problems, the friction force microscope (FFM) [12-16] and the scanning surface potential microscope (SSPM) [18-20] have recently proven useful for imaging and identifying unlike chemical surfaces such as the partially film-covered surfaces [12,15,20] and the domains of phase-separated mixed monolayers of hydrocarbon (HC) and fluorocarbon (FC) amphiphiles [13-19].

Although friction and surface potentials are highly material dependent, another physical property that is more sensitive to surface chemical properties is desired for definitive chemical assignment. The spectroscopic information is the most useful for this purpose. The lateral resolution of conventional microscopic spectroscopy [21] is, however, limited by the diffraction of the observed light in the same way as those of the fluorescence microscopy and the Brewster angle microscopy described above.

On the other hand, the resolution of SNOM [22] is determined by the dimensions of the microscopic light source (or detector) and the separation between the probe and the sample surface rather than the diffraction limit. As the light source, a microlithographic hole [23], a small aperture in a metal film [24,25], a micropipette [26,27], a capillary tube containing a fluorescence dye at the tip end [28], and an optical fiber tip [29-31] have been used. For the separation control, the steep decrease in a tunneling current [32], a shear force [33,34], or the intensity of an evanescent wave [29-31] with an increase in the separation was utilized.

Although the study of the SNOM has a long history [22], the optical images have not been taken with a high resolution until recent years. One of the most progressed SNOMs

*M. Nieto-Vesperinas and N. García (eds.), Optics at the Nanometer Scale 205–221.*

uses the shear force to control the distance and an aluminum-coated tapered optical fiber [35]. The fluorescence microscopy based the above designs was extended to imaging single molecules [36,37] and single proteins [38] and the determination of molecular orientation [36] and fluorescence lifetimes [39-41]. The other is the combined SNOM and AFM in the contact mode based on a silicon nitride cantilever with a pyramidal tip [42]. In the latter microscope, the tip was used simultaneously as the SNOM optical probe and the AFM tip.

Recently, we took advantage of non-contact AFM to control the tip-sample distance for SNOM without mechanical damages [43-45]. The combined non-contact AFM and SNOM was demonstrated to be used as scanning near-field fluorescence microscopy (SNFM) and fluorescence spectroscopy for localized micro areas [46]. In the present review, various application of this method to fluorescence microscopy and spectroscopy of organic thin films will be described.

## 2. SNOM-AFM for Fluorescence Microscopy and Spectroscopy

An optical fiber tip sharpened and bent by heating was mounted on a stainless steel cantilever as shown in Fig. 1(a) at the beginning [43,44,46]. Later, a bent optical fiber itself was used as a cantilever [45,47-50] as shown in Fig. 1(c). The distance between the tip and the sample was controlled by an AFM operation with a laser beam deflection [51,52] in which the cantilever was vibrated at the resonance frequency. The decrease in an amplitude of the vibrating cantilever by the van der Waals force [53,54] was used to control the distance with a piezoelectric scanner.

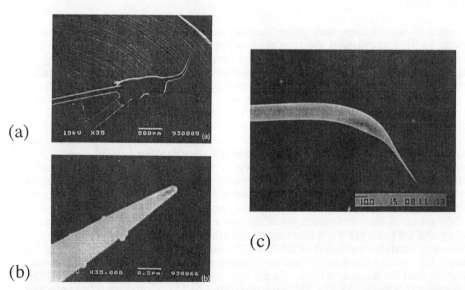

Figure 1. Scanning electron microscope (SEM) images of (a) a sharpened and bent optical fiber mounted at the end of a stainless steel cantilever (2 mm length) [44], (b) an etched optical fiber tip, the diameter of which was approximately 100 nm [44], and (c) an Al coated optical fiber tip for modified SNOM-AFM [45]. The bent optical fiber itself with a tip (c) was used as an AFM cantilever and a part of the optical fiber was polished and Al coated as a mirror for laser beam deflection [45,47].

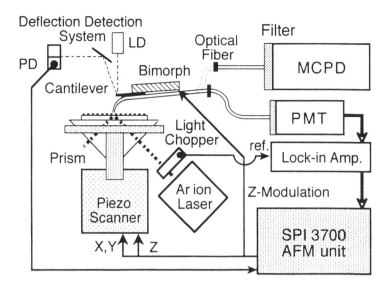

Figure 2. The schematic diagram of the initial SNOM-AFM for fluorescence imaging and fluorescence spectroscopy for micro-area [46].

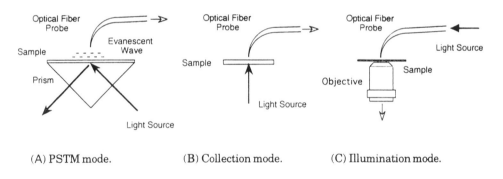

(A) PSTM mode.          (B) Collection mode.          (C) Illumination mode.

Figure 3. Schematics of typical optical modes for SNOM-AFM [44,48].

Figure 2 shows the schematic diagram of our initial SNOM-AFM for SNFM and fluorescence spectroscopy for micro-areas[43,44,46]. The uncoated optical fiber tip picks up an evanescent wave of an Ar ion laser beam for excitation and fluorescence light from the irradiated sample surface on a prism. The laser beam was modulated by a light chopper and impinged on a sample-prism interface at the incidence angle of 45°. The picked-up light was introduced into a photomultiplier tube (PMT) through the optical fiber for SNOM. For recording a fluorescence spectrum, the optical fiber was introduced into a slit of an Otsuka Electronics IMUC-7000 intensified multichannel photodetector. Selection of the excitation or the fluorescence light was done by optical filters placed in front of the PMT window or the slit of IMUC-7000. Scanning of the tip and imaging of lateral distribution of the intensity of the evanescent wave or the fluorescence light were carried out with a Seiko Instruments SPI-3700 AFM unit which contains a non-contact AFM function and accepts a user's input signal. A lock-in amplifier was used to distinguish a modulated optical signal from optical noise.

Since samples were irradiated by the Ar ion laser continuously during SNFM and fluorescence spectroscopy in the initial SNOM-AFM shown in Fig. 2 (mode (A) in Fig. 3), photodamage of the samples was apparent. To reduce the photodamage, samples were irradiated by light from an Al coated optical fiber tip (mode (C) in Fig. 3) which was again sharpened and bent by heating with $CO_2$ laser (Fig. 1(c)) [45,48-50]. Mode (B) in Fig. 3, i.e. collection mode, was used, for example, to map the distribution of light intensity at one end of the cut optical fiber when the laser beam was introduced at the other end of the optical fiber [47].

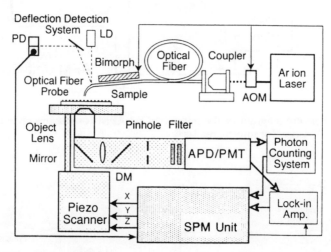

Figure 4. The schematic diagram of the modified SNOM-AFM in which a light was irradiated by an optical fiber probe to a sample surface and transmitted light or fluorescence is collected by an objective lens through the sample and introduced to a photo detector [48].

In the modified SNOM-AFM for SNFM, fluorescence was collected with a microscope objective as shown in Fig. 4. Since an optical filter emitted fluorescence of the filter itself upon the Ar ion laser irradiation, a dichroic mirror without fluorescence emission was placed in front of the filter to eliminate the Ar ion laser excitation beam [48]. A plate with a pinhole was also placed to reduce stray light.

## 3. Fluorescence Microscopy and Spectroscopy of Microlithographically Patterned Samples

Chromium checker-board patterns with various dimensions on a quartz substrate gifted by Dai Nippon Printing Co., Saitama were used as test samples for optical absorption of the Ar ion excitation beam and for the non-contact AFM [43,45-47]. For the fluorescence micro-scopy and specrtoscopy, the chromium patterns were coated by spin coating with a thin polyvinyl alcohol film containing fluorescein [46]. The first example of SNOM and SNFM of the patterned samples was demonstrated by the initial SNOM-AFM as described below.

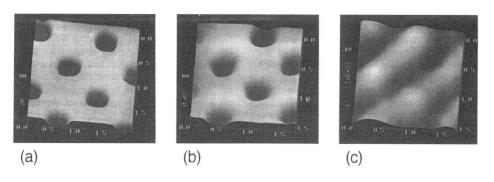

Figure 5. AFM topographic images of a chromium checkerboard pattern (ca. 25 nm in height) by the use of (a) a microfabricated Si3N4 tip in contact AFM mode and (b) an optical fiber tip in non-contact AFM mode. (c) A map of the evanescent light intensity picked up by the tip on the same surface observed simultaneously with the topographic image shown in (b) [46].

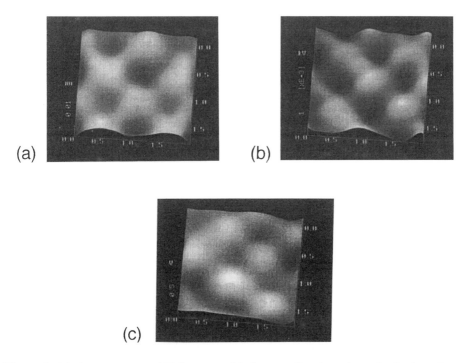

Figure 6. (a) A non-contact AFM topographic image of a chromium checkerboard pattern coated with a polyvinyl alcohol thin film containing fluorescein recorded simultaneously with (b) a map of the total light intensity picked up by the optical fiber tip. (c) A map of the fluorescence intensity on the same sample but on a different location [46].

As shown in Fig. 5(a), the chromium checker-board pattern (ca. 25 nm in height) without the fluorescent film was imaged with a high resolution less than 10 nm by the use of an ordinary microfabricated silicon nitride tip (Olympus) in the contact AFM mode. The non-contact AFM topographic image of the same sample but probably the different location was taken by the use of the optical fiber tip with a resolution of ~ 100 nm (Fig. 5(b)). The map of the evanescent light intensity picked up by the tip gave the corresponding checker-board pattern with a resolution of ~ 100 nm as shown in Fig. 5(c). It is clear from comparison between the non-contact topographic AFM (Fig. 5(b)) and the light intensity image recorded simultaneously (Fig. 5(c)) that the light intensity was high on the chromium uncovered parts.

The non-contact AFM topographic images (2 x 2 μm²) of the dye-film coated sample is shown in Figs. 6(a). The topographic height difference between the chromium covered and the uncovered parts was diminished by the film coating but the chromium covered parts are likely to be still higher (ca. 7 nm) than the uncovered parts. The map of the total light intensity measured simultaneously on the dye-film covered sample is also shown in Fig. 6(b). Here, the total light consists of the evanescent wave of Ar ion laser transmitted through the chromium pattern with the dye film, the fluorescence emitted by the dye, and other optical noise including the scattered light of the laser diode for the AFM beam deflection. The last component could, however, be removed by the use of the lock-in amplifier for the SNOM. It is clear from comparison between two images in Figs. 6(a) and 6(b) that the lower light intensity areas in Fig. 6(b) correspond to the topographically higher areas in Fig. 6(a) where chromium seems to cover the quartz substrate. This agrees with the above observation on the sample without the dye film that the intensity of the evanescent wave behind the opaque chromium film is weak.

The screening of the excitation beam by the chromium pattern also gave clear contrast in the fluorescence intensity mapping as shown in Fig. 6(c), where the excitation beam picked up by the optical fiber was cut by the filter in front of the PMT (Fig. 1). The fluorescence intensity map had, however, to be recorded within a few minutes because appreciable photodamage and photobleaching of the dye were seen. As described above, a light irradiation mode through the optical fiber tip (Fig. 3(C)) is preferable to avoid this damage.

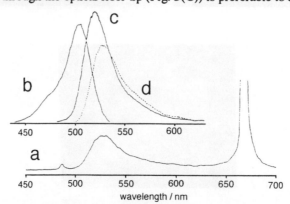

Figure 7. (a) A fluorescence spectrum recorded by positioning the optical fiber tip above the chromium uncovered area of the sample surface shown in Fig. 6(c). (b) A fluorescence excitation spectrum (emission wavelength, 570 nm), (c) an emission spectrum (excitation wavelength, 470 nm), and (d) an emission spectrum (dotted line) (excitation wavelength, 488 nm) with a cut filter SC-52. The spectra (b) - (d) were observed with a large area sample by the use of an ordinary Hitachi 850 fluorescence spectrophotometer [46].

Figure 7(a) shows a fluorescence spectrum recorded with an Otsuka Electronics IMUC-000 intensified multichannel photodetector by positioning the tip above the chromium ncovered area. A fluorescence band of fluorescein appearing around 525 nm was in good greement with that observed with a large area sample by the use of an ordinary Hitachi 50 fluorescence spectrophotometer with the same optical filter (Fig. 7(d)). In addition to e fluorescence band, in Fig. 7(a), a peak at 488 nm due to Ar ion laser passing through e cut filter is seen together with a strong peak at 670 nm due to the laser diode for the FM beam deflection.

## . Scanning Near-field Fluorescence Microscopy (SNFM) of Fluorescent Polystyrene Spheres with Modified SNOM-AFM

By using the modified SNOM-AFM (Fig. 4), we investigated test samples that were ated by spin coating with a thin polyvinyl alcohol film containing fluorescent and non-uorescent polystyrene latex spheres on a cover glass [48]. The spheres with 100 nm ameter were doped with fluorescent dye which has an absorption maximum at 490 nm d a fluorescence maximum at 515 nm. The non-fluorescent spheres with 200 nm in ameter were also utilized. Figures 8(a) and 8(b) show, respectively, a 2.7 x 2.7 $\mu m^2$ FM topographic image and a 2.7 x 2.7 $\mu m^2$ SNOM fluorescence image, observed multaneously, of the same sample surface containing the polystyrene spheres.

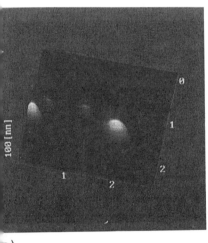 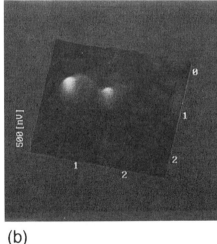

a)                                    (b)

gure 8. (a) A 2.7 x 2.7 $\mu m^2$ non-contact AFM topography image recorded simultaneous-with (b) a 2.7 x 2.7 $\mu m^2$ map of the fluorescence light intensity of a sample containing ge non-fluorescent (200 nm) and small fluorescent (100 nm) polystyrene spheres [48].

The fluorescent measurement was performed by the use of SNOM in illumination mode ig. 3(C) and Fig. 4). As shown in Fig. 4, the Ar ion laser beam was modulated and upled into the optical fiber probe by a coupler for the illumination of the sample surface the probe. The modulated incident laser beam by an acoustic optical modulator was nchronized with the vibration of the optical fiber probe for illuminating the sample face only when the optical tip end was located closely to the surface.

The dichroic mirror (DM510, Olympus) reduced transmitted excitation laser beam (488 nm) and, a Y-51 Toshiba filter and an LS-550 Corion filter cut a 488 nm excitation light and a 680 nm light due to the AFM beam deflection laser diode, respectively. The input 488 nm Ar ion laser power was about 50 mW at the laser output and the fluorescence light detected by the avalanche photodiode were about 2000 counts per second. When the dichroic mirror and the filters were removed, the transmitted and collected light by the objective lens was about $10^6$ counts per second detected by the avalanche photodiode. The light power was depends on the probe used. The images of the fluorescent spheres have been measured previously with shear force regulation [35]. Here, we obtained fluorescent spheres image by a non-contact AFM in combination with SNOM.

The contrast difference between the Figs. 8(a) and 8(b) was attributable to the existence of the fluorescent dye in polystyrene spheres. It is clear from comparison between two images that the higher fluorescence light intensity areas in Fig. 8(b) correspond to the smaller fluorescent polystyrene spheres in Fig. 8(a), while, no fluorescence light was observed for the larger non-fluorescent polystyrene spheres.

The images in Fig. 8 were taken within 5 min, but the fluorescence intensity was not decreased during several repetitive imagings. By the use of the SNOM in the illumination mode in the fluorescence measurements, illumination of the excitation light through the optical fiber tip is weak enough to diminish the photobleaching in comparison with the above results under illumination by a total reflection mode. The optical resolution was improved by the modulated incident laser beam synchronized with the vibration of the optical fiber probe for illuminating the sample surface only when the optical tip end was located closely to the surface. This was because the vibration of the cantilever had to be kept large (50 - 100 nm) to improve the AFM control in the modified SNOM-AFM. If the vibration becomes smaller (< 10 nm), there will be no need to synchronize with an incident laser beam and a continuous wave will be better for the increase in collected photons.

The polystyrene latex spheres in Fig. 8(a) seem to have larger diameters than their actual size probably because the probe size not only in horizontal dimensions but also in vertical dimensions like a cone may affect the image size. The heights of the polystyrene spheres in Fig. 8(a) showed values expected from the bead diameters. If the exact shape of a sample is known, we will be able to obtained the probe shape from a corresponding AFM topography image.

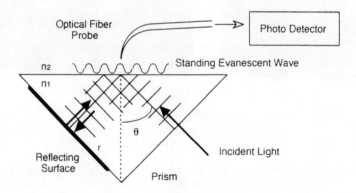

Figure 9. The schematic diagram of the modified SNOM-AFM for collecting photons from a standing evanescent wave bound to the surface of a prism which has a reflecting coating with aluminum film on output face [48].

The lateral resolution of the microscope was examined by measuring a standing evanescent wave, which was formed by reflecting the incident laser beam on a prism [55,56]. Figure 9 shows a schematic diagram of the modified SNOM-AFM for collecting photons from a standing evanescent wave bound to the surface of the prism. Two counter propagating evanescent waves are formed by total reflection of an Ar ion laser beam inside the 90° prism which coated with a reflecting Al-film on an output face, and interfere each other and result in a standing wave. In this setup, the modulation period, $\Lambda$, depends on the wavelength ($\lambda_0$ = 488 nm), the angle of the incidence ($\theta$ = 45°), and the refractive index change at the surface ($n_1$ = 1.51, $n_2$ = 1.0), and is about 228 nm. The probe-sample separation was kept constant by non-contact AFM control during the scan of the probe on the prism.

Figure 10(a) shows a 2.0 x 0.8 $\mu m^2$ image of light intensity of the evanescent wave on the prism surface which was taken by the modified SNOM-AFM setup shown in Figs. 4 and 9. Figure 10(b) shows a line section of the evanescent field which was shown in Fig. 10(a). For the probe aperture size estimation, we used an approximate method proposed by Meixner et al. [55,56]. According to their analysis [55], there is a relation between the amplitude of the measured signal Aexp and an effective aperture radius $\Delta$: Aexp = exp [-$(\pi\Delta/\Lambda)^2$]. The experimental value for Aexp which is defined as the ratio of the peak-to-peak power to the average power was determined to be 0.39 from the line section in Figure 10(b) Thus, the effective aperture radius of the probe end was estimated to be about 70 nm from the above equation taking account of $\Lambda = \lambda_0/(2n_1\sin\theta)$ [55]. This estimation is good standard for the effective aperture size of the SNOM fiber probe.

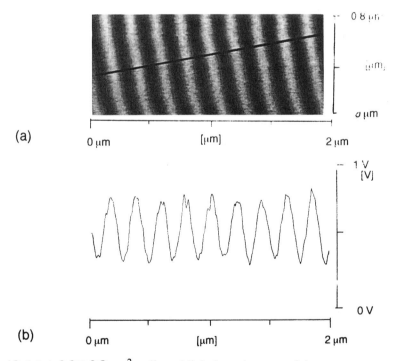

Figure 10. (a) A 2.0 x 0.8 $\mu m^2$ collected light intensity map of the evanescent wave on the prism surface. (b) A line section of the intensity of the evanescent field along a line in (a).

## 5. SNFM of Preferential Dissolution of Fluorescent Cyanine Dye with Two Long Alkyl Chains into HC Phase in A Phase-separated HC-FC Mixed Monolayer

A small amount of a cationic cyanine dye with two long alkyl chains (CD) was added [49] to see whether this HC fluorescence dye is preferentially dissolved in the HC island phase in phase-separated mixed monolayers of HC and FC amphiphiles [13-17]. Size of islands studied by AFM and FFM changed from place to place and ranged from 2 to 4 μm in diameter, but on average the size became larger than that of the two component stearic acid (SA) - FC amphiphile (PFECA) mixture [13,14] by addition of 1/40 CD in the SA - PFECA (1:1) mixture. Friction in the FC sea again was found to be higher than that observed in the HC islands.

When CD concentration was decreased from 1/40 to 1/100 in the SA-PFECA-CD mixed monolayers, the size of islands was decreased on average. A typical cyclic contact (i.e. tapping) AFM and a SNFM image of the mixture simultaneously recorded by the modified SNOM-AFM are shown in Figs. 11(a) and 11(b), respectively [49]. For stable distance control for AFM of this monolayer sample, the cyclic contact mode was better than the non-contact mode. In the AFM image obtained by cyclic contact mode, topographic image between the islands and the sea is reversed in contrast with those obtained by AFM in contact mode. Although difference in elastic properties between these two areas may be responsible to the reversal, origin of the change in contrast has not been clarified yet and is now under intensive study. The SNFM image in Fig. 11(b) clearly shows that the modified SNOM-AFM can be used to observe fluorescence images of islands less than 0.3 μm. In other words, the SNOM-AFM can be used to irradiate a single island domain. The image also confirms the preferential dissolution of CD in the HC islands. It is concluded from these observations that the cationic A-S-D triad for charge separation will be preferentially dissolved in islands consisting of light harvesting pyrene derivatives in artificial photosynthetic molecular devices [57,58].

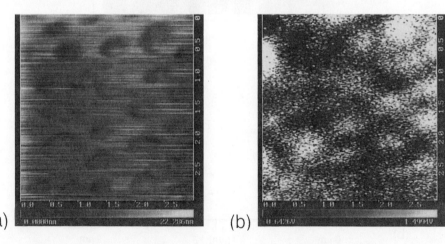

(a)  (b)

Figure 11. (a) A cyclic contact AFM and (b) a fluorescence micrograph (SNFM image) of a mixed monolayer of SA-PFECA-CD (1:1:1/100) polyion complexed with PVA deposited on a cover glass plate measured simultaneously with the modified SNOM-AFM [49].

## 6. Spectroscopy of Micro-area for Study of Degradation Mechanism of Organic Electroluminescence (EL) Devices

As an application of the modified SNOM-AFM to fluorescence spectroscopy of micro-area for practical use, we have studied degradation mechanism of organic EL devices [50]. Since dramatic improvement in organic EL devices with layered organic thin films reported by Tang and Vanslyke [59], EL devices with ITO / TPD / Alq3 / Metal structures have been studied extensively. However, even in this type of device, formation of dark spots is found to be one of factors limiting the lifetimes of such devices. In addition, mechanism of the dark spot formation has not been clarified yet.

We recently found that organic layers in the EL device were mixed each other by molecular diffusion [60,61] upon heating or Joule heating by current flow through the EL device. Mixing by diffusion was concluded by measurements of the change in fluorescence spectra of the TPD / Alq3 layered films by heating. However, the fluorescence spectra were measured over large areas and we could not correlate the fluorescence observation to the dark spot formation. All we have known so far is that Alq3 diffusing into the TPD layer became energy acceptor of excited TPD and blue fluorescence from TPD was quenched to give green fluorescence from Alq3. Fluorescence microscopy of micro-area, especially dark spots, with the modified SNOM-AFM was expected to give some answer to this question. However, in the following experiment, strict near-field conditions were avoided, because not only overlaid Alq3 (50 nm) but also underlaid TPD layer (50 nm) should be illuminated in the same way as in an ordinary far field method for comparison with the fluorescence spectra obtained with a conventional spectrophotometer. To confirm the conditions, aperture of about 10 $\mu$m was used. The large aperture was still less than the size of the dark spots and gave high sensitivity for fluorescence spectroscopy. During the fluorescence spectral measurements, the tip-sample separation was kept to be about 1 $\mu$m by retracting the z-piezo by switching off the photodiode for AFM after positioning the tip to an micro-area of interest by non-contact AFM mode.

Figure 12(a), 12(b), and 12(c) show a typical EL, a fluorescence, and an optical micrograph, respectively, of an ITO / TPD / Alq3 / Al EL cell degraded by EL operation. All micrographs were taken with a conventional fluorescence microscope. The EL micrograph (a) was taken on an EL cell at 28 min after applying 23 without UV or visible light illumination. The fluorescence (b) and the optical (c) micrograph were taken on the same area as that shown in (a) after switching off the applied voltage for EL with UV (355 nm) and visible light (a halogen lamp) illumination, respectively [50,62]. All images were taken from the ITO side.

In Fig. 12(a), a large number of dark spots are seen. As shown in Fig. 12 (b), colors on the fluorescence image were greenish-blue and yellowish-green in the normal and the dark spot area, respectively. The greenish-blue emission was observed all over the EL device before applying voltage. By heating other EL device prepared separately for 30 min at 80 °C, it was found that the homogeneous yellowish-green fluorescence was emitted all over the device as reported previously [60,61]. Black spots in the fluorescence image located in the centers of the dark spots appeared as bright spots in the optical image in Fig. 12(c). This indicates that the centers of the dark spots were too heavily damaged to emit luminescence and Al on the centers was oxidized to transparent Al2O3 [62].

Figure 13 shows fluorescence spectra of an EL device with dark spots by positioning the tip above (a) the normal surface, (b) the outer ring of the dark spot, and (c) the center of the dark spot. The top Al electrode was removed by a scotch tape for better fluorescence spectroscopy [63]. As shown in Fig. 13(c), the fluorescence intensity at 520 nm due to Alq3 increased at the expense of the fluorescence intensity at 400 nm for TPD at the center of the dark spot. The change in the fluorescence spectra can be interpreted in terms of

216

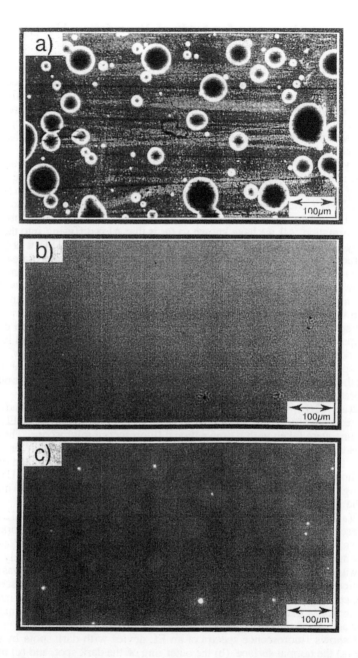

Figure 12. (a) A typical EL, (b) a fluorescence, and (c) an optical image of a degraded EL device. The EL image (a) was taken at 28 min after applying 23 V. The fluorescence (b) and the optical (c) image were taken for the same area as in (a) after switching off 23 V [50].

energy transfer in the mixed film formed by the interdiffusion in the same way as observed by heating described above.

It is concluded from these observation together with fluorescence microscopy of model films prepared by co-deposition that the presence of diffused Alq₃ less than 1 % in the TPD layer quenched TPD fluorescence [50]. In addition, the change in electrical properties of the organic thin films by mixing indicated that one of main factors to form the dark spots is this interdiffusion by Joule heating of the EL device under operation [64].

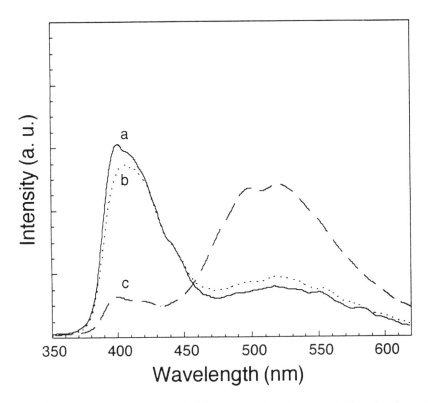

Figure 13. Fluorescence spectra recorded by positioning the optical fiber tip above (a) the normal surface, (b) the outer ring of the dark spot, and (c) the center of the dark spot [50].

## 7. Conclusions

The scanning fluorescence microscopy and fluorescence spectroscopy of the localized area of various samples were successfully performed with a high resolution of ~ 100 nm by our SNOM-AFM. By the distance regulation with AFM in the non-contact mode, the SNOM-AFM will be applicable to soft samples like biological materials without serious mechanical damages. For further improvement of lateral resolution of the SNOM-AFM, we are studying presently photon energy confinement by taking advantage of energy transfer in a well designed optical fiber tip illustrated in Fig. 14 in the same way as an antenna system of a natural photosynthetic device [57, 58,65].

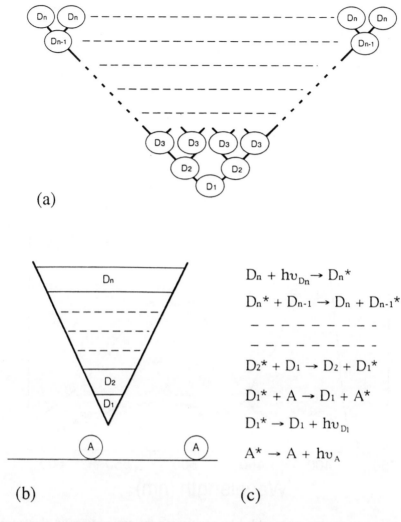

Figure 14. Confinement of photon into nano-tips by (a) a synthetic and (b) a coating method by taking advantage of short-distance interaction of energy transfer (c).

## Acknowledgements

I would like to thank my colleagues in our laboratory, N. Yamamoto, H. Monobe, A. Koike, L. M. Do, and R. Sawada and in Seiko Instruments, H. Muramatsu, N. Chiba, K. Nakajima, N. Homma, and T. Ataka who collaborated in this work. This work was partially supported by a Grant-in-Aid for Scientific Research on New Program (03NP0301) from Ministry of Education, Science, and Culture of Japan and Special Coordination Funds of the Science and Technology Agency of the Japanese Government.

# References

1. Lösche, M. & H. Möhwald, H. *Rev. Sci. Instrum.*, **55**, 1968 (1984).
2. Weis, R. M. & McConnell, H. M. *Nature* **310**, 47 (1984).
3. Fujihira, M., Nishiyama, K., Hamaguchi, Y. & Tatsu, Y. *Chem. Lett.*, **1987**, 253 (1987).
4. Hönig, D. & Möbius, D. *J. Phys. Chem.*, **95**, 4590 (1991).
5. Henon, S. & Meunier, J. *Rev. Sci. Instrum.*, **62**, 936 (1991).
6. Ries, Jr., H. E. & Kimball, W. A. *J. Phys. Chem.*, **59**, 94 (1955); *Nature*, **181**, 901(1958).
7. Fryer, J. R., Hann, R. A. & Eyres, B. L. *Nature*, **313**, 382 (1985).
8. Barraud, A., Leloup, J., Maire, P. & Ruaudel-Teixier, A. *Thin Solid Films*, **133**, 133 (1985).
9. Smith, D. P. E., Bryant, A., Quate, C. F., Rabe, J. P., Gerber, C. & Swalen, J. D. *Proc. Natl. Acad. Sci. USA*, **84**, 969 (1987).
10. Meyer, E., Howald, L., Overney, R. M., Heinzelmann, H., Frommer, J., Güntherodt, H.-J., Wagner, T., Schier, H. & Roth, S. *Nature, 349*, 398 (1991).
11. Bourdieu, L., Silberzan, P. & Chatenay, D. *Phys. Rev. Lett.*, **67**, 2029 (1991).
12. Meyer, E., Overney, R., Brodbeck, D., Howald, L., Lüthi, R., Frommer, J. & Güntherodt, H.-J. *Phys. Rev. Lett.*, **69**, 1777 (1992).
13. Overney, R. M., Meyer, E., Frommer, J., Brodbeck, D., Lüthi, R., Howald, L., Güntherodt, H.-J., Fujihira, M., Takano, H. & Gotoh, Y. *Nature, 359*, 133 (1992).
14. Meyer, E., Overney, R., Lüthi, R., Brodbeck, D., Howald, L., Frommer, J. Güntherodt, H.-J., Wolter, O., Fujihira, M., Takano H. & Gotoh, Y. *Thin Solid Films, 220*, 132 (1992).
15. Fujihira, M. & Morita, Y. *J. Vac. Sci. Technol.*, **B12**, 1609 (1994).
16. Frommer, J. *Angew, Chem., Int. Ed.*, **31**, 1298 (1992).
17. Fujihira, M. In *Forces in Scanning Probe Methods*: Güntherodt, H.-J., Anselmetti, D., Meyer, E., Eds., Kluwer Academic Publishers: NATO ASI Series E: Applied Sciences, Vol. 286, p. 567, 1995.
18. Fujihira, M., Kawate, H. & Yasutake, M. *Chem. Lett.*, **1992**, 2223 (1992).
19. Fujihira, M. & Kawate, H. *Thin Solid Films*, **242**, 163 (1994).
20. Fujihira, M. & Kawate, H. *J. Vac. Sci. Technol.*, **B12**, 1604 (1994).
21. *New Techniques of Optical Microscopy and Microspectroscopy*, Cherry, R. J., Ed., In *Topics in Molecular and Structural Biology*, Vol. 15, Macmillan Press: Hampshire, 1991.
22. See e.g., *Near Field Optics*, Pohl, D. W., Courjon, D., Eds., NATO ASI Series, Kluwer: Dordrecht, 1993, and references therein.
23. Lewis, A., Isaacson, M., Harootunian, A. & Murray, A. *Ultramicroscopy*, **13**, 227 (1984).
24. Pohl, D. W., Denk, W. & Lanz, M. *Appl. Phys. Lett.*, **44**, 651 (1984).
25. Fischer, U.Ch. *J. Vac. Sci. Technol.*, **B3**, 386 (1985).
26. Harootunian, A., Betzig, E., Isaacson, M. & Lewis, A. *Appl. Phys. Lett.*, **49**, 647 (1986).
27. Betzig, E., Lewis, A., Harootunian, M., Isaacson, M. & Kratschmer, E. *Biophys. J.*, **49**, 269 (1986).
28. Lieberman, K., Harush, S., Lewis, A. & Kopelman, R. *Science*, **247**, 59 (1990).
29. Courjon, D., Sarayeddine, K. & Spajer, M. *Opt. Commun.*, **71**, 23 (1989).
30. Reddik, R. C., Warmack, R. J. & Ferrell, T. L. *Phys. Rev.*, **B39**, 767 (1989).
31. De Fornel, F., Goudonnet, J. P., Salomon, L. & Lesniewska, E. *Proc. SPIE*, **1139**, 77 (1989).

220

32. Dürig, U., Pohl, D. W. & Rohner, F. *J. Appl. Phys.*, **59**, 3318 (1986).
33. Betzig, E., Finn, P. L., Weiner, J. S. *Appl. Phys. Lett.*, **60**, 2484 (1992).
34. Toledo-Crow, R., Yang, P. C., Chen, Y. & Vaez-Iravani, M. *Appl. Phys. Lett.*, **60**, 2957 (1992).
35. Betzig, E. & Trautman, J. K. *Science*, **257**, 189 (1992).
36. Betzig, E. & Chichester, R. J. *Science*, **262**, 1422 (1993).
37. Tarrach, G. Bopp, M. A., Zeisel, D. & Meixner, A. J. *Rev. Sci. Instrum.*, **66**, 3569 (1995).
38. Dunn, R. C., Allen, E. V., Joyce, S. A., Anderson, G. A. & Xie, X. S. *Ultramicroscopy*, **57**, 113 (1995).
39. Dunn, R. C., Holtom, G. R., Mets, L. & Xie, X. S. *J. Phys. Chem.*, **98**, 3094 (1994).
40. Xie, X. S. & Dunn. R. C. *Science*, **265**, 361 (1994).
41. Ambrose, W. P., Goodwin, P. M., Martin, J. C. & Keller, R. A. *Proc. SPIE*, **2125**, 2 (1994).
42. Van Hulst, N. F., Moers, M. H. P., Noordman, O. F. J., Faulkner, T., Segerink, F. B., Van der Werf, K. O., De Grooth, B. G. & Bölger, B. *Proc. SPIE*, **1639**, 36 (1992).
43. Fujihira, M., Monobe, H., Muramatsu, H. & Ataka, T. *Chem. Lett.*, **1994**, 657 (1994).
44. Muramatsu, H., Chiba, N., Ataka, T., Monobe, H. & Fujihira, M. *Ultramicroscopy*, **57**, 141 (1995).
45. Muramatsu, H., Chiba, N., Homma, K., Nakajima, K., Ataka, T., Ohta, S., Kusumi, A., & Fujihira, M. *Appl. Phys. Lett.*, **66**, 3245 (1995).
46. Fujihira, M., Monobe, H., Muramatsu, H. & Ataka, T. *Ultramicroscopy*, **57**, 118 (1995).
47. Chiba, N., Muramatsu, H., Ataka, T. & Fujihira, M. *J. J. Appl. Phys.*, **34**, 321 (1995).
48. Fujihira, M., Monobe, H., Yamamoto, N., Muramatsu, H., Chiba, N., Nakajima, K. & Ataka, T. *Ultramicroscopy*, in press.
49. Fujihira, M., Sakomura, M., Aoki, D. & Koike, A. *Thin Solid Films*, in press.
50. Fujihira, M., Do, L. M., Koike, A. & Han, E. M. *Appl. Phys. Lett.*, in press.
51. Meyer, G. & Amer, N. M. *Appl. Phys. Lett.*, **53**, 1045, 2400 (1988).
52. Alexander, S., Hellemans, L., Marti, O., Schneir, J., Elings, V., Hansma, P. K., Longmire, M. & Gurley, J. *J. Appl. Phys.*, **65**, 164 (1989).
53. Martin, Y., Williams, C. C. & Wickramasinghe, H. K. *J. Appl. Phys.*, **61**, 4723 (1987).
54. McClelland, G. M., Erlandsson, R., Chiang, S. in: *Review of Progress in Quantitative Nondestractive Evaluation*, Thompson, D. O., Chimenti, D. E., Eds., Vol. 6B, Plenum: New York, 1987, p. 307.
55. Meixner, A. J., Bopp, M. A. & Tarrach, G. *Appl. Opt.*, **33**, 7995 (1994).
56. Meixner, A. J., Zeisel, D., Bopp, M. A. & Tarrach, G. *Opt. Eng.*, **34**, 2324 (1995).
57. Fujihira, M. *Advances in Chemistry Series*, **240**, 373 (1994).
58. Fujihira, M. In *Thin Films*, Ulman, A., Ed., Vol. 20, Academic Press: Boston, p. 239, 1995.
59. Tang, C. W. & Vanslyke, S. A. *Appl. Phys. Lett.*, **51**, 913 (1987).
60. Han, E. M., Do, L. M., Yamamoto, N. & Fujihira, M. *Chem. Lett.*, **1995**, 57 (1995).
61. Han, E. M., Do, L. M., Yamamoto, N. & Fujihira, M. *Thin Solid Films*, in press.
62. Do, L. M., Han, E. M., Niidome, Y., Fujihira, M., Kanno, T., Yoshida, S., Maeda, A. & Ikushima, A. J. *J. Appl. Phys.*, **76**, 5118 (1994).

63. Do, L. M., Oyamada, N., Han, E. M., Yamamoto, N. & Fujihira, M. *Thin Solid Films*, in press.
64. Do, L. M., Han, E. M., Koike, A., Yamamoto, N. & Fujihira, M. to be submitted.
65. Fujihira, M. *Mat. Res. Soc. Symp. Proc.*, **227**, 47 (1992).

# FLUORESCENCE LIFETIME CONTRAST COMBINED WITH PROBE MICROSCOPY

O. H. WILLEMSEN, O.F.J. NOORDMAN, F.B. SEGERINK
A.G.T. RUITER, M.H.P. MOERS & N.F. VAN HULST
*Applied Optics group, Faculty of Applied Physics*
*& MESA Research Institute, University of Twente,*
*P.O.Box 217, 7500AE Enschede, the Netherlands.*

ABSTRACT.    Fluorescence lifetime imaging is combined with atomic force microscopy in an integrated scanning microscope using a silicon-nitride probe. The time decay of fluorescence is measured at each image position with a resolution of 50 ps by time correlated single photon counting using a frequency doubled mode-locked Ti-Sapphire laser and fast electronics. Images of a mixture of fluorescence labelled latex spheres are presented where the lifetime contrast of the different spheres can be directly correlated to the topography as detected by force microscopy.

## 1. Introduction

In Fluorescence Lifetime-resolved Imaging Microscopy (FLIM) [1] the temporal charactcristics of luminescent emission in a fluorescence microscope image are recorded at each position in the image field [2]. Thus it is possible to measure directly the local mean lifetime of the emission and to discriminate among several fluorescent components in the image based on their distinct decay times.

The time decay of fluorescence is sensitive for environmental factors, which is mainly due to the lifetime of the excited state, typically 10 ns, long enough to involve processes with other molecules over 10 nm away at the moment of excitation. Thus photodynamic processes as molecular rotation, solvent and matrix relaxation, chemical reactions, quenching processes, singlet-triplet intersystem crossing, complex formation or energy transfer can all be studied best by their characteristic time dependence. Obviously mapping of the heterogeneity of the molecular environment in e.g. biological specimens is performed with higher selectivity by combining both temporal and spectral information.

The dynamic response of fluorophores can be measured using frequency domain or time domain methods. In the frequency domain the excitation is RF modulated (10 - 100 Mhz) and the phase shifted demodulated fluorescence response is measured [2, 3]. This phase fluorometry is a fast and efficient way to determine the mean fluorescence

*M. Nieto-Vesperinas and N. García (eds.), Optics at the Nanometer Scale 223–233.*

decay time, however for low signal levels approaching photon counting the method is not suitable. In the time domain the complete fluorescence decay curve is measured after excitation with a sub-ns pulse. Generally the time profile is determined using photon counting techniques and sampling over many excitation pulses. The acquisition of full decay curves over a total image is very slow, therefore often a compromise is chosen: instead of measuring the whole temporal profile the fluorescence is integrated over two pre-set time windows and the ratio of the integrated intensities is related to an average lifetime [4].

Currently the application of confocal FLIM arrangements is developing fast, especially in the biological domain. Yet ultimately diffraction sets a lower limit to the spatial resolution that can be obtained. Further improvement of resolution, while maintaining temporal optical contrast, requires measurement in the optical near-field using probe microscopy methods. Also the technique of Near-field Scanning Optical Microscopy (NSOM) has developed fast over recent years [5], including observation and spectroscopy of single molecules [6, 7]. Recently Dunn et al. [8] have demonstrated fluorescent lifetime measurements on photosynthetic membranes and Sunney Xie et al. [9] and Ambrose et al. [10] showed time resolved images of single molecules.

In this paper we present recent progress using an integrated set-up [11, 12] with simultaneous fluorescence lifetime imaging and atomic force microscopy.

## 2. Fluorescence lifetime

Upon irradiation with light close to the resonance for the singlet $S_0 \Rightarrow S_1$ transition molecules are excited within about 10 fs to one of the vibrational state of $S_1$. The processes competing with the subsequent fluorescence are inter-system crossing (ISC) to the triplet manifold, internal conversion (IC) to the ground state and dissociation (D) through a photochemical reaction. Vibrational relaxation to the lowest vibrational level of $S_1$ occurs on ps time scale and is generally complete before electronic relaxation. The subsequent radiative decay rate $k_R$ depends on the overlap of the molecular orbitals for the electron in $S_0$ and $S_1$. The value of $k_R$ extends from $10^9$ s$^{-1}$ for fully allowed transitions ($\pi\pi^*$ in common dye molecules) to $10^7$ s$^{-1}$ for symmetry forbidden transitions (aromatic molecules) down to $10^5$ s$^{-1}$ ($n\pi^*$ , e.g. carbonyl compounds). Due to the competing decay channels (ISC, IC, D) the effective fluorescence decay time $\tau_F$ is shorter than $k_R^{-1}$,

$$\tau_F = (k_R + k_{ISC} + k_{IC} + k_D)^{-1}$$

The observed fluorescence decays exponentially with the effective decay time $\tau_F$. However due to the existence of several decay channels in practise often a more complex (bi-exponential) decay is observed. The quantum yield $\Phi_F$ for fluorescence is $k_R \cdot \tau_F < 1$, typically 0.5 - 0.9 . Thus it is clear that, even without interaction with the environment, the amount of fluorescence depends on the size of the radiative rate constant $k_R$ relative to the sum of all rates ($\Sigma k$). For optimum signal in a fluorescence experiment fluorophores with dominant radiative decay ($10^8$ - $10^9$ s$^{-1}$ ) and quantum yield close to unity have to be chosen.

## 3. Time correlated single photon counting

Temporal profiles of the fluorescence decay are recorded using Time Correlated Single-Photon Counting (TCSPC) [13, 14]. The basic electronic scheme is given in figure 1.

Optical excitation pulses are generated from a mode-locked Ti:Sapphire laser (Spectra-Physics, Tsunami 3960), which gives 80 fs pulses in the wavelength region 700-1100 nm at a repetition rate of 82 MHz. After frequency doubling pulses in the region 350-550 nm are delivered every 12.2 ns. A sample is illuminate and the generated fluorescence is detected with a fast sensitive detector: a MicroChannel Plate PhotoMultiplier Tube (MCP-PMT, Eldy EMI-132/300), which allows detection of single photons and gives reproducible electric pulses with a width of 420 ps. Part of the direct laser pulse is detected by a fast photo-diode (BPX65). Both MCP signal and photo-diode signal are amplified (EG&G Ortec VT120 & 9306) and discriminated. Hereto constant fraction discriminators (EG&G Ortec 935) are used, which enable exact timing of the pulse maximum with an accuracy only limited by the jitter and noise of the pulse. The discriminated pulses act as start and stop pulses for a Time-to-Amplitude Converter (TAC, EG&G Ortec 566). The central part of the TAC is a capacitor that starts to charge up at the moment the start pulse arrives and ends charging up by the arrival of a stop pulse. Thus a pulse is generated at the output of the TAC with an amplitude proportional to the time difference between start and stop pulse. The pulses are stored in a multichannel analyzer. Every laser shot that generates a fluorescent photocount starts the TAC and gives an output pulse. With repetitive pulses a histogram of the time difference distribution builds up. Of course TCSPC is based on zero or one photon per laser pulse. As a consequence the fluorescent photon emission rate is limited to ~ 1% of the laser repetition rate, i.e. $10^6$ photons/s. In practise the maximum count rate is further limited to $5 \cdot 10^4$ counts/s by the A/D conversion in the multi-channel analyzer.

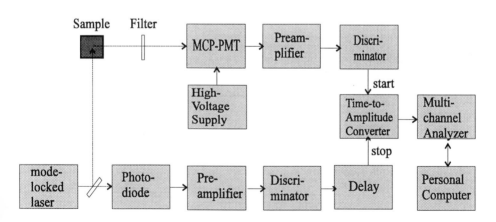

*Figure 1: Detection electronics for Time Correlated Single Photon Counting.*

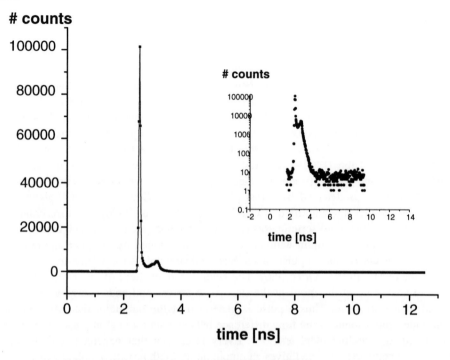

*Figure 2: System response function with FWHM = 70 ps; inset log-scale.*

The measured temporal intensity profile $I_F(t)$ will be a convolution of the actual profile $G(t)$ and the system response function $P(t)$, as given by

$$I_F(t) = \int_0^t P(t')G(t-t')dt' \ .$$

If the instrument response function is sufficiently narrow compared to the fluorescence lifetime no deconvolution is necessary. The instrumental function is determined by the combination of laser, detector and electronics and has been tested for its impulse response, as plotted in figure 2. The system response function has a peak of 70 ps FWHM and 140 ps at the 10% level. The main peak is followed by a broader second peak with a 22× lower signal at 610 ps delay, caused by the discriminator action.

Due to the 12.2 ns repetition period of the laser and partial overlap of the start pulse with previous stop pulses in practise a time window of 8 ns is available. The maximum lifetime to be determined is about 4 ns if one requires two decades of fluorescence decay for a sufficient exponential fit.

## 4. FLIM - AFM set-up

The microscopic set-up is a modified version of a combined NSOM/AFM [15] as described by Ruiter *et al.* [16]. The sample is scanned using a large area 200 x 200 μm piezo-electric scanner. A SiN integrated probe and an optical beam deflection system constitute the AFM part. Two large working distance objectives (0.55 NA) are arranged confocally with two pinholes to form the FLIM part. The ∅ 1 μm focus on the sample area is ~ 4 μm shifted relative to the AFM probe. An additional Hg-lamp and CCD camera facilitate alignment of AFM and FLIM and allow direct viewing of the sample area of interest. At each x-y position of the scanner 3000 photocounts are collected in a histogram of 64 bins, yielding a time profile for each image pixel.

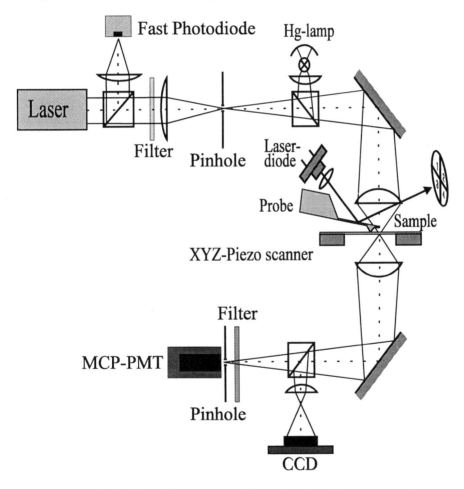

*Figure 3: Combined Fluorescence Lifetime Imaging Microscope and Atomic Force Microscope (FLIM - AFM).*

## 5. Measurements and Discussion

### 5.1. MEASUREMENTS ON FLUORESCENT MOLECULES IN SOLUTION.

The fluorescent molecules Streptavidin PE (Becton & Dickinson, Streptavidin PhcoErythyn) and F18 (fluorescein isothiocyanate, FITC, coupled to 18 carbon atoms), both in PBS (Phosphate Buffer Saline) at pH = 7.37, were used for lifetime measurement. The molecules are excited at 485 nm and 460 nm, respectively. Figure 4 shows a typical fluorescence time profile for F18. The exponential decay shows some oscillations caused by the electronics. The decay curves are fit to a single exponential decay, taking into account an offset due to the darkcounts of the MCP. Comparison of the determined lifetimes to literature values in table 1 shows good agreement.

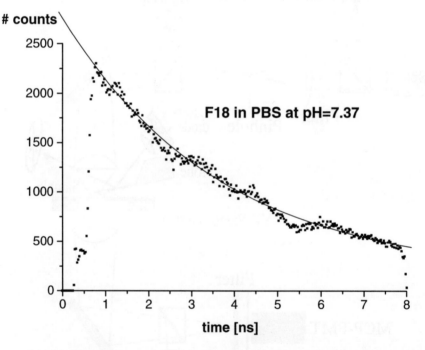

*Figure 4: Fluorescence decay of F18 in PBS at pH=7.37 with 460 nm excitation.*

| Fluorophore | Exp., with / without offset [ns] | Literature [ns] |
|---|---|---|
| Streptavidin PE | 3.2 / 3.3 | 2.9 / 3.2 [17] |
| FITC/F18 | 3.7 / 4.3 | 4.2 [18], 3.8 [19] |

*Table 1: Experimental an literature fluorecence lifetimes.*

## 5.2. MEASUREMENTS ON FLUORESCENTLY STAINED BEADS

A mixture of two kinds of fluorescence labelled latex spheres was prepared differing both in size and time-resolved behaviour was prepared: 70% 1.0 μm beads stained with fluorescein and 30% 1.8 μm beads (both Polysciences). The fluorescence decay of the separate spheres was measured, both in water and on glass. Figure 5 shows decay curves, where the profile of the 1.8 μm beads is multiplied by 3 for clarity. The 1.0 μm beads display a single exponential decay with 2.4 ns lifetime. The 1.8 μm beads show a fast and a slow component in the decay, with 600 ps and 3.0 ns lifetime, respectively, while the average single exponential decay time is 1.6 ns.

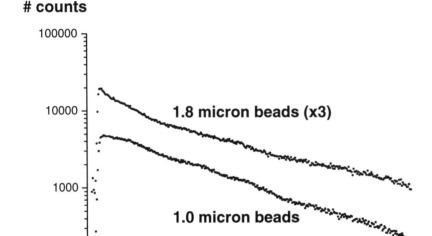

*Figure 5: Fluorescence decay 1.8 μm and 1.0 μm labelled beads*

## 5.3. COMBINED FLUORESCENCE LIFETIME AND FORCE MICROSCOPY.

For microscopic imaging a droplet of the mixed beads was put on an object glass and slightly heated to fix the spheres to the glass. Thus monolayers and multilayers of closely packed mixed spheres are fabricated. The sample is illuminated at 480 nm of the frequency doubled Ti:Sapphire laser to excite both fluorophores. While scanning the sample time profiles and AFM signal are recorded simultaneously for each pixel. The offset AFM probe and laser focus have sufficient overlap within a 30 x 30 μm scan area to allow comparison. Images at different time windows were extracted from the data and plotted together with the AFM scan in figures 6 and 7.

230

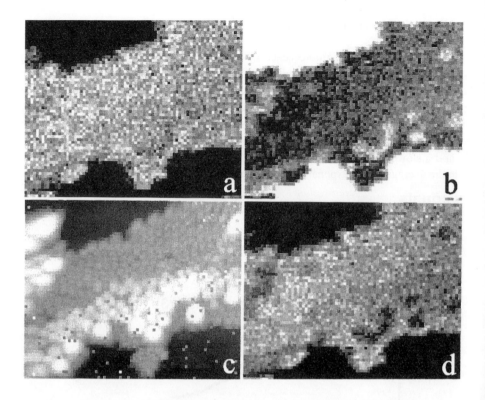

*Figure 6:*

*30 x 22 μm image of a mixture of two types of fluorescence labeled
latex spheres on glass: 1.8 μm and 1.0 μm:*

*(a) Fluorescence signal in the first 400 ps after excitation.*
*(b) Fluorescence signal in 3.1 - 7.0 ns interval after excitation.*
*(c) Simultaneously recorded atomic force image.*
*(d) Ratio image of fast over slow fluorescence.*

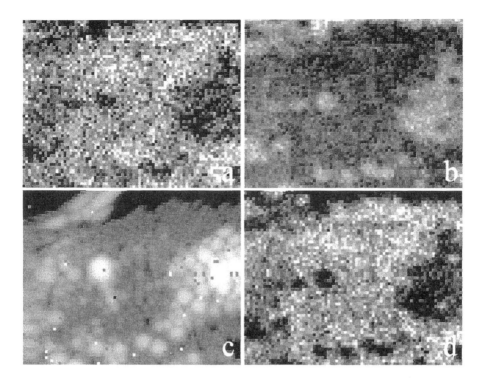

*Figure 7:*

*30 x 22 µm image of a mixture of two types of fluorescence labeled*
*latex spheres on glass: 1.8 µm and 1.0 µm:*
*(a) Fluorescence signal in the first 780 ps after excitation.*
*(b) Fluorescence signal in 3.1 - 7.0 ns interval after excitation.*
*(c) Simultaneously recorded atomic force image.*
*(d) Ratio image of fast over slow fluorescence.*

232

In figures 6 and 7 image (a) shows the fast fluorescence signal in the first few 100 ps, while image (b) shows the long term fluorescence in the 3.1 - 7.0 ns time interval. Comparison clearly shows the contrast reversal: the small spheres dominate in the first ns and the slow component of the larger beads dominates on the longer timescale of several ns. It should be noted that at each pixel 3000 counts are stored and consequently the images show the relative contrast for the given time interval. The corresponding topographic AFM images in (c) show the glass substrate with monolayers and sometimes double layers of 70% small beads and 30% large beads. Finally in (d) images are presented of the ratio of fast over slow fluorescence, which show enhanced contrast compared to the time window images. The discrimination of the spheres on size in the AFM images corresponds well to the time contrast in the FLIM images. In figure 6c a monolayer to double layer step of small beads is observed while the lifetime is unchanged in figure 6d. On an additional layer with large beads the lifetime does change, as expected. Some structures in the AFM image are non-fluorescent and do not show up in the optical images.

## 6. Conclusions

A combined FLIM - AFM set-up has been presented. Fluorescence lifetime images have been obtained simultaneously with topographic AFM images. The time resolution is determined by a system response function of 70 ps FWHM. Lifetimes of 1-4 ns can be determined. Lifetimes of 3.3 and 3.7 ns have been determined for Streptavidin PE and F18, respectively. The spatial resolution of the optical system is about 600 nm, as determined by the numerical aperture of the objectives. The lateral resolution of the AFM is determined by tip sharpness and drift during time integration for fluorecence detection. Different fluorescent beads have been discriminated on fluorescence lifetime characteristics in good correspondence with the discrimination on size.

The time window for detection can be extended using a pulse picker for the mode-locked laser, which allows measurement of longer lifetimes.

The optical resolution will be further improved in a near-field optical arrangement using a metal-coated fibre probe or an integrated NSOM probe [16]. Thus the nano-environment of fluorophores will be analyzed by the photo-dynamic response.

## 7. Acknowledgements

The authors thank Thomas Widén, Patrik Reimer, Kees van der Werf, John van Noort, Bart de Grooth and Cees Otto for their contributions and suggestions. This research is mainly supported by the Dutch Foundation for Fundamental Research (FOM, division Atomic Physics & Quantum Electronics) and the European network on Near-field Optics and Nanotechnology.

# 8. References

1. Gadella, T.W., Jovin, T.M. and Clegg, R.M. (1993) Fluorescence lifetime imaging microscopy (FLIM): spatial resolution of microstructures on the ns time scale, *Bioph. Chem.* **48**, 221-239.

2. Lakowicz, J.R. and Berndt, K.W. (1991) Lifetime-selective fluorescence imaging using an RF phase sensitive camera, *Rev. Sci. Instrum.* **62**, 1727-1734.

3. Laczko, G., Gryczynski, I., Gryczynski, Z., Wiczk, W., Malak, H. and Lakowicz, J.R. (1990) Phase fluorometry. *Rev. Sci. Instrum.* **61**, 2331.

4. Buurman, E.P., Sanders, R., Draaijer, A., Gerritsen, H.C., van Veen, J.J.F., Houpt, P.M. and Levine, Y.K. (1992) Fluorescence lifetime imaging using a confocal laser scanning microscope, *Scanning* **14**, 155-159.

5. Betzig, E. and Trautman, J.K. (1992) Near-field optics: microscopy, spectroscopy and surface modification beyond the diffraction limit, *Science* **257**, 189-195.

6. Betzig, E. and Chichester, R.J. (1993) Single Molecule observed by Near field Scanning Optical Microscopy, *Science* **262**, 1422-1425.

7. Trautman, J.K., Macklin, J.J., Brus, L.E. and Betzig, E. (1994) Near field spectroscopy of single molecules at room temperature, *Nature* **369**, 40-42.

8. Dunn, R.C., Holtom, G.R., Mets, L. and Sunney Xie, X. (1994), Near-field fluorescence imaging and fluorescence lifetime measurement of light harvesting complexes in intact photosynthetic membranes, *J. Phys. Chem.* **98**, 3094-3098.

9. Sunney Xie, X. and Dunn, R.C. (1994) Probing Single Molecule Dynamics, *Science* **265**, 361-364.

10. Ambrose, W.P., Goodwin, P.M., Martin, J.C. and Keller, R.A. (1994) Alterations of Single Molecule Fluorescence Lifetimes in Near-Field Optical Microscopy, *Science* **265**, 364-367.

11. Van Hulst, N.F., Moers, M.H.P., Noordman, O.F.J., Faulkner, T., Segerink, F.B., van der Werf, K.O., De Grooth B.G. and Bölger B. (1992) Operation of a Scanning Near field Optical Microscope in reflection in combination with a scanning force microscope, *SPIE* **1639**, 36-43.

12. Moers, M.H.P., Tack, R.G., Van Hulst, N.F. and Bölger, B. (1994) Photon scanning tunneling microscope in combination with a force microscope, *J. Appl. Phys.* **75**, 1254-1257.

13. O.Connor, D.V. and Philips, D. (1984) Time correlated single photon counting, Academic Press, London,

14. Wilkerson, C.W. Goodwin, P.M., Ambrose, W.P., Martin, J.C. and Keller, R.A. (1993) Detection and lifetime measurement of single molecules in flowing sample streams by laser-induced fluorescence, *Appl. Phys. Lett.* **62**, 2030-2032.

15. Van Hulst, N.F., Moers, M.H.P. and Bölger, B. (1993) Near-field optical microscopy in transmission and reflection modes in combination with force microscopy, *J. Microscopy.* **171**, 95-105.

16. Ruiter, A.G.T., Moers, M.H.P., Jalocha, A. and van Hulst, N.F. (1995) Development of an integrated NSOM probe. *Ultramicroscopy* **58**, in press.

17. Grabowski, J. and Gantt, E. (1978) *Photochemistry and Photobiology* **28**, 39.

18. Chen, R.F.(1969) *Archives of Biochemistry and Biophysics* **133**, 263

19. Haughland, R.P. (1992) Handbook of fluorescent probes and research chemicals, 5th ed., Molecular Probes Inc. Eugenen, Oregon.

# TOWARDS SNIM:

# SCANNING NEAR-FIELD MICROSCOPY IN THE INFRARED

F. KEILMANN
*Max-Planck-Institut für Biochemie*
*82152 Martinsried, Germany*

**ABSTRACT.** The SNOM has successfully replaced the wavelength by an aperture in determining an optical microscope's ultimate resolution. Thus it appears now feasible that also long-wavelength infrared radiation is exploited for microscopy. The expected benefit is primarily an extension of available contrast mechanisms, especially to include "fingerprint" vibrational absorption specific to the infrared which can identify an object's chemical composition.

Of the problems to attain a high-resolution SNIM we discuss specifically the cutoff and skin depth effects of metal-coated lightguides. We find that proper choices of tip materials and tip geometries can fully circumvent both obstacles, resulting in the prediction that the SNIM will challenge the SNOM's 20 nm spatial resolution.

Furthermore we show that the SNIM's resolution should be improvable even beyond 20 nm by implementing antenna tips as already demonstrated in "apertureless" SNOMs. As an extreme example of going beyond the wavelength limit, we report preliminary scale experiments with an apertureless radiowave-SNIM which has already resolved sub-micrometer features.

## 1. Introduction

In scanning near-field optical microscopes (SNOMs) a metallized fiber tip confines the light such that about $\lambda/10$ resolution is routinely achieved [1-3]. This represents an improvement in resolving power by about one order of magnitude compared to classical microscopy, and therefore, many new possibilities arise for surface and thin film analysis in physics, chemistry, biology and medicine.

A great number of microscopic contrast mechanisms can be exploited in SNOMs, in principle as many as there are spectroscopic techniques which involve photons. Up to now mainly absorption, phase shift and fluorescence [4] have been demonstrated.

*M. Nieto-Vesperinas and N. García (eds.), Optics at the Nanometer Scale 235–245.*

Nearly no attempt has been made to vary the wavelength outside the visible region, especially, to take advantage of infrared, microwave and radio waves. In these spectral regions specific types of contrast, such as conductivity or vibrational absorption wait to be exploited for practical use [5]. Ironically it was early radio wave experiments [6] that first demonstrated $\lambda/30$ confinement of radiation by employing metal apertures, and thus lead to the SNOM development [6].

## 2. Local Spectroscopies

In the zoo of modern scanning microscopies (SXM, where X stands for the sensed quantity such as tunnel current or force), the common aim is to obtain local material information. The most important one is the material topography, i.e. the location of the sample´s surface, for the simple technical reason that the sensing tip must move very closely along the sample surface in order that high spatial resolution can be achieved.

Further information requires that more quantities be measured at a given location. This can already be realized by employing a dual or multifunction sensor, for example one which can sense tunnel current and force simultaneously. The more common approach however has been to employ a local spectroscopy. This term means that at each sensor position a series of data is recorded before moving to the next pixel, such as a force vs. distance relation, or a current vs. bias relation. In the same way that the latter has been able to yield information on energies of surface states one should of course expect interesting information from classical electromagnetic spectroscopy. The experimental challenge is therefore to perform high-resolution imaging offered by the SNOM technique with a broadband visible or infrared spectrum.

## 3. Infrared Excitations

In general all condensed materials exhibit electronic excitation in some spectral region which thus can serve as the basis for spectroscopic material characterization. In the case of metallic binding the materials are highly conductive which means that electronic excitation is possible for all frequencies from zero to about visible light. The resulting infrared reflectivity is near 100% which makes it hard to characterize metals spectroscopically in this region.

Non-conducting materials, however, can be electronically excited only at about visible or higher frequencies. This means that insulators either appear coloured such as a dye solution or a ZnSe crystal and thus exhibit a colour contrast, or they are visibly transparent like water or diamond and exhibit no colour contrast. Thus, although the large group of transparent materials possess specific electronic resonances these can not be exploited for microscopic material characterization. This is the reason why taking the trouble of artificial staining of specimens is so widespread in biological and medical microscopy work.

A second general class of excitation found in all materials is the relative motion of bound atoms in form of lattice or molecular vibrational modes. Even for light atoms with strong bonds such as in the HF molecule the vibrational frequencies are below the visible range. For this reason vibrational absorption contrast has not been exploited in optical microscopy.

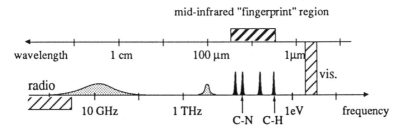

*Figure 1.* Schematic view of vibrational absorption resonances in the mid and far infrared spectral regions, and of the microwave absorption of water.

The infrared absorption bands are commonly recorded with infrared spectrometers and play an established rôle in analytical chemistry. These spectra routinely serve to identify visibly colourless materials from inorganic compounds to biopolymers. An atlas of "fingerprint" absorption spectra is available to identify even mixtures of materials in a quantitative manner. Some vibrational frequencies are shown in Fig. 1. Clearly the mid-infrared region, i.e. the decade of wavelengths from about 3 to 30 μm, is the center of interest in analytical spectroscopy. Both broad-band infrared sources and detectors exist in this region, and therefore can probably be adapted to serve in future SNIM systems.

Further vibrational modes and other low-energy excitations of considerable interest to solid-state physics exist at even longer, far-infrared wavelengths. These resonances are however not important for the characterization of most materials, partly because they are greatly broadened if not overdamped by collisional disturbance at room temperature, and also, because far-infrared sources and detectors perform less well. Only when we go to even longer wavelengths we find that at radio frequencies polar molecules such as water start to respond, by orienting in the alternating radio field, and thus give rise to absorption spectra which can be used for their identification.

238

## 4. Materials for Infrared Tips

An extension of the typical SNOM´s optical elements to the mid infrared mainly requires to find a suitable focusing waveguide equivalent to a sharpened monomode optical fiber. The $SiO_2$- or polymer-based optical fibers are not usable because of strong vibrational absorption. A group of crystalline materials exists with full transparency in the mid-infrared. Common members are diamond, Si, Ge, ZnSe and KBr. The latter is used to fabricate multimode infrared waveguides, but tip sharpening methods comparable to glass pulling or glass etching have not yet been developed. Most probably SNIM tips will have to rely on hybrid optics, combining very different elements for first guiding the light near to the tip, and then to focus and confine the radiation through a small aperture.

Metal coatings should be well suited to confine infrared radiation because of their high reflectance. Caution should be payed to the increase of penetration or skin depth which after the classical electrodynamics theory [7] increases with wavelength according to

$$\delta = (4\pi\mu\nu\sigma)^{-1/2} \; , \tag{1}$$

where $\mu$ is the magnetic permeability, $\nu$ the frequency, and $\sigma$ the conductivity. With the value of Au, $\sigma = 4.5 \times 10^5 \; (\Omega cm)^{-1}$, we obtain $\delta = 1.2 \; \mu m$ at $\nu = 1$ GHz. Therefore it is predicted that at a typical microwave frequency of $\nu = 10$ GHz a sub-micrometer resolution can just be achieved. If one were to consider the use of superconducting apertures an even better confinement by at least an order of magnitude is possible for such long-wavelength radiation [8,9].

At higher frequencies the simple relationship (1) ceases to be valid. The modification arises because the quantized electronic excitation across electronic bands comes into play. A common way to take these influences into account has been to replace in (1)

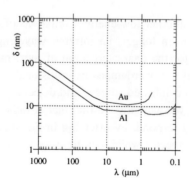

Figure 2.    Power penetration depth $\delta$ vs. wavelength of visible and infrared radiation.

the value of the conductivity $\sigma$ by a spectral conductivity function $\sigma(\nu)$ which deviates from a constant value only at infrared and higher frequencies. When we use tabulated values of $\sigma(\nu)$ for well-ordered metals [10] we obtain from (1) the curves in Fig. 2. The result is quite enlightning: we observe nearly no difference in $\delta$ for a very broad spectral range extending from the optical down to most of the mid-infrared region. Therefore, the aperture-limited microscope should perform equally well over all this broad range. With this prediction it should be even more desirable to start building instrumentation to extend the SNOM principle into the infrared.

## 5. Geometrical Principles to Overcome the Cutoff Rejection

The metallic aperture should not only confine the radiation but at the same time permit a high radiation throughput of the microscope. In this latter sense the SNOMs have up to now performed very badly: the loss in transmitted light power has been extremely high, of the order of 50 db equivalent to 99.999% [4]. This loss comes from the use of a short cutoff waveguide section between the aperture and either the light source or the light detector.

The cutoff effect follows from Maxwell´s equations solved for the interior of a hollow tube with conducting walls. The result is that all waves below a so-called cutoff frequency $\nu_c$ propagate as evanescent modes, i.e. they are strongly reflected when a tapered hollow metal waveguide is designed to concentrate an electromagnetic wave beyond a diameter of about $\lambda/2$. This is unfortunately what happens when a metal-coated glass fiber is tapered down to about $\lambda/10$, as in the usual SNOM. The cutoff frequency is given by $\nu_c = c\,(\zeta a)^{-1}$ where c is the light velocity, a is the maximum diameter of the metal tube, and $\zeta$ is a value near 2 depending only on the tube cross sectional form. The power transmittance of a length L of an (infinitely well conducting) cutoff waveguide is $T = \exp(-2\gamma L)$ where $\gamma = 2\pi\,(\nu_c^2 - \nu^2)^{1/2}\,/c$. While this may be modified for non-perfect conductivity [11] the essence of the cutoff effect for microscopy is that while the loss may just be bearable in the SNOM it would totally ruin signal detection in a SNIM, because of a larger effective L.

Fortunately non-cutoff waveguides do exist and are well-known in electronics. Although their use for SNOMs has been suggested some time ago [5,12] such structures have up to now not been realized in the visible or infrared regions of interest here, only at somewhat longer wavelengths [13]. In radio or microwave electronics, on the other hand, a virtually lossless coupling to sub-wavelength structures is routinely realized with non-cutoff waveguides. To illustrate this consider the coaxial cable which is a hollow metal waveguide complemented by a central metal wire. Since the two parts are electrically not connected the coaxial guide can completely confine the electromagnetic field of propagating waves at arbitrarily long wavelengths, down to static fields. Because of this, the requirement of a high throughput presents no problem for a radiowave microscope, and all efforts to realize similar structures for the use of infrared and visible wavelengths should pay off as decisive factors to improve the SNOM and make the SNIM possible.

## 6.    Geometrical Principles to Overcome the Penetration Depth Limit

Let us consider in more detail how the penetration of light in the aperture material limits the spatial resolution of a SNOM and SNIM. Even when we now assume a non-cutoff metal waveguide such as the coaxial guide, a metallic aperture consisting of the outer conductor has to confine the radiation. The confinement is then again limited by the power penetration depth $\delta$ of light into the metal [2]. The confinement should not be better than about $2\delta$ even when the aperture diameter is tapered down to near zero. This would predict that a radio-SNOM at 1 GHz corerponding to a wavelength of 30 cm can not resolve features sizes below about 2 $\mu$m.

Fortunately this argument is not valid. While we have considered radiation inside the metal, we have neglected what happens at the metal surface. The change in dielectric values from the transparent material to the metal produces however a strong electric field step at the interface. With curved surfaces as near the tip apex this leads to a strong field concentration effect. Therefore we can expect that a metal tip — even if it conducted so badly that $\delta$ were larger than the curvature radius R ! — would still function to concentrate the field to a dimension R in its immediate neighborhood. An approximate simple description of the field concentration by a metal tip is provided by electrostatics. Its applicability should extend to infrared and visible radiation provided the tip dimension remains $<<\lambda$ . While more detailed simulations could be performed with electromagnetic field programs developed for electrical engineering, the electrostatic model seems an appropriate intuitive guideline when it comes to improve the SNOM´s resolution limit beyond 20 nm.

Important experiments have already used the field concentration effect near a small metallic or also dielectric "antenna" [14-16], and indeed have demonstrated an order of magnitude improved resolution. This class of "apertureless" SNOMs will be the choice when the highest resolution is sought, although problems of suppressing the large straylight background require special attention. A way to improve the coupling between a waveguide mode and a metal tip antenna, by employing propagating surface plasmons, has recently been conjectured in our laboratory [17].

## 7. Antenna Tips for Coaxial-Guide SNIMs

The impedance of a coaxial line is $Z = 60 \ \epsilon^{1/2} \ \ln(d_o/d_i)$ $[\Omega]$ where $d_i$ is the diameter of the inner conductor, and $d_o$ is the inner diameter of the outer conductor. By tapering such a line a coaxial tip can be formed (Fig. 3). If this taper is made such that the ratio $d_o/d_i$ and thus $Z$ stays constant, any reflection along the taper can be avoided. At the tip, however, the line ending represents a strong mismatch so that nearly complete reflection results. The fraction of power transmitted into free space is of the order of $(d_o/\lambda)^2$, for small protrusion lengths $s \ll d_o$. Note, however, that the purpose of the coaxial tip is not at all to radiate into free space, but rather to provide a highly confined field.

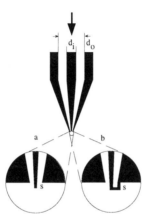

*Figure 3.* Cross-section of cylindrical coaxial metal waveguide, tapered for focusing electromagnetic waves, and terminated to probe either electric (a) or magnetic (b) interactions.

The termination can be either open or shorted. This choice allows to define the electromagnetic field in the proximity of the tip to be either nearly electric or nearly magnetic. As the sample is placed in this region one can choose thereby whether contrast is to be obtained from electronic interactions (e.g. carrier mobility or dielectric polarization) or from magnetic interactions (e.g. nuclear or electronic spin resonance). The latter geometry (Fig. 3b) of creating a magnetic near-field has recently been applied by us to excite electron spin transitions in semiconductors, as will be published elsewhere. In this letter we restrict the discussion on the *electric* near-field antenna (Fig. 3a).

The influence of a sample in close proximity of the terminated coaxial line is most easily measured by its perturbation of the reflected wave as has been used for a scanning near-field microwave microscope by Fee *et al.* [5]. A pure phase contrast was obtained from small non-absorbing (dielectric and metallic) objects in close proximity to the inner conductor tip. The size and sign of the phase change was explained by simply modelling the propagation along the object and back as a prolongation of the waveguide. This view of course requires that the wave couples to the object with good match. Generally, the sensitivity of reflectometry relies on the stability of the source because the sample provides a small change to a large return signal. The discrimination against the background becomes an even more severe problem when reflection occurs along the taper part of the line.

## 8.    Radio-Wave Transmission Microscopy

We have pursued a "transmission mode" of operating a radiowave microscope: we detect the wave which passes from the sensing aperture through the sample region into a receiving waveguide. Other than with the SNOM the sample can be in the near fields of both the sending and the receiving antennas. The interaction can be derived from the electrostatic field distribution. The observed contrast is determined only by the amount by which the sample alters, in amplitude and phase, the coupling between transmitting and receiving near-field antennas.

Our experiment [18] uses two coaxial cables in close proximity, one connected to a radiowave source, the other to a receiver. The inner conductors protrude by a length s equal to $d_0 \approx 2$ mm (Fig. 3a). The upper one is sharpened to a tip with radius of about 6 μm (Fig. 4a). As a source we use a signal generator (HP 8657A) which can be tuned between 1 MHz and 1 GHz. As a receiver we use a vector voltmeter (HP 8508A). The transmittance is found to decrease monotonically with the distance between the tips. The samples are two versions of a commercial polarizer foil (Buckbee-Mears Co.) consisting of 40 μm thick mylar with evaporated thin ( 1 μm) gold stripes. To define a separation between the tip and the sample we added a 6 μm mylar foil as a spacer. The sample was slightly pressed between the tips and scanned using a motor-driven micrometer stage.

Figure 4 shows examples of the transmittance scan traces. Both samples produce a modulation such that the transmittance is minimal when the tip is centered over a metal stripe. The image profile is reminiscent of the reflectometer micrograph of a similar object (Fig. 2b in ref. [5]). For the coarser sample (Fig. 4b) the power transmittance is modulated by about 1 db, i.e. by about 20 %. The phase is modulated by 2 degrees. The finer sample produces a similar radiowave image, but with the modulation reduced by a factor 2 to 3.

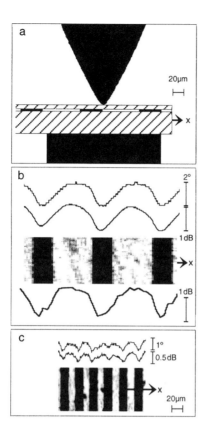

*Figure 4.* Scanning radiowave transmission microscope, with
(a) sending tip (top), sample (middle) and detecting tip (bottom);
(b) images for two frequencies, 740 MHz (above) and 5 MHz
(below); (c) image of 25 μm period sample at 740 MHz.

Nearly identical scanned images are obtained also using widely different radio frequencies. Comparison of the upper and lower traces of Fig. 4b demonstrates that the image taken with a hundred times longer wavelength is identical, i.e. also in its absolute contrast. This proves directly the electrostatic model, since the wavelength is no longer a relevant quantity to describe the image formation.

We thus have found that the radiowave microscope resolves structures of about 10 μm size although the waveguide aperture $d_0$ is very much larger. Clearly it is the near field around the sharp tip that matters. Ongoing experiments in our laboratory with finer tips made of W have begun to show that the radiowave transmission microscope can resolve submicrometric features, well below the penetration depth in W [19].

## 9.     Equivalence of rf-SNOM and Capacitance Microscopy

The electrostatic coupling model [18] leads us to bridge the gap between the SNOM and the published scanning capacitance microscope (SCM) [20-23]. SCMs map the capacity between a scanning metal tip and a metal sample. The capacitance is measured by its tuning of a high-Q resonant circuit. Scanning reveals information on locally varying separation or dielectric constant, or their combination. Very high spatial resolution has already be attained. A problem with SCMs (reminiscent of a similar one with the above mentioned "reflectance" radio microscope [5]) is the high background signal provided by stray capacitance: the total capacitance measured has only a very small contribution from the spherical tip apex, but is dominated by the capacitance originating from the conical needle leading to the tip. The latter is 100 pF [22] or even 3000 pF [21]. With suitable shielding this value was reduced to 1 fF [23]. In [23] a tip radius of 50 nm was achieved by etching. Theoretically, an R = 50 nm spherical conductor in 50 nm distance from a conductive plane represents a capacitance of 0.01 fF. If such a small capacitance could be resolved on the stray background an accordingly high spatial resolution could be achieved. This is true for any microscopy based on capacitance, especially also for any near-field microscopy, be it at high frequency, radio or even optical frequency. While the penetration depth does not limit the resolution as discussed above it does however limit the shielding quality of the outer conductor, i.e. the minimum extent of stray fields.

## 10.    Conclusion

We have discussed that a SNIM infrared microscope should be built because it has important applications as a "chemical" microscope. The resolution can be well in the sub-micrometric domain.

To this aim, rather than trying to extend the somewhat non-ideal techniques of the SNOM to longer wavelengths, we find it a better way to extend radiowave concepts such as non-cutoff guides, tip antennas and heterodyne receivers to shorter wavelengths.

## 11.    Acknowledgements

It is a pleasure to acknowledge cooperation with D.v.d. Weide (Newark),
R. Guckenberger and A. Kramer (Martinsried).

## 12. References

[1] E.H. Synge, Philos. Mag. **6**, 356 (1928)

[2] D.W. Pohl, W. Denk and M. Lanz, Appl. Phys. Lett. **44**, 651 (1984)

[3] D.W. Pohl, Europhys. News **26**, 75 (1995)

[4] H.F. Hess, E. Betzig, T.D. Harris, L.N. Pfeiffer and K.W. West, Science **264**, 1740 (1994)

[5] M. Fee, S. Chu and T. Hänsch, Opt. Commun. **69**, 219 (1989)

[6] E.A. Ash and G, Nicholls, Nature **237**, 510 (1972)

[7] J.D. Jackson, Classical Electrodynamics, Wiley 1962

[8] M.P. Tu, K. Mbaye, L. Wartski and J. Halbritter, J. Appl. Phys. **63**, 4586 (1988)

[9] J. Halbritter, J. Supercond. **5**, 171 (1992)

[10] H.J. Hagemann, W. Gudat and C. Kunz, DESY report SR-74/7 (1974)

[11] F. Keilmann, Int. J. Infrared and Millimeter Waves **2**, 259 (1981)

[12] F. Keilmann, U.S. patent no. 4,994,818 (1991, filed 1988)

[13] F. Keilmann, Infrared Phys. Technol. **36**, 217 (1995)

[14] J. Wessel, J. Opt. Soc. Am. **B2**, 1538 (1985)

[15] M. Specht, J.D. Pedarnig, W.M. Heckl and T.W. Hänsch, Phys. Rev. Lett. **68**, 476 (1992)

[16] F. Zenhausern, Y. Martin and H.K. Wickramasinghe, Science **269**, 1083 (1995)

[17] F. Keilmann and R. Guckenberger, Deutsche Patentanmeldung 19522546 v. 21.6.1995

[18] F. Keilmann, D.W. van der Weide, T. Eickelkamp, R. Merz and D. Stöckle, to be published

[19] A. Kramer, F. Keilmann and R. Guckenberger, in preparation

[20] J.R. Matey and J. Blanc, J. Appl. Phys. **57**, 1437 (1985)

[21] C.D. Bugg and P.J. King, J. Phys. E **21**, 147 (1988)

[22] C.C. Williams, W.P. Hough and S.A. Rishton, Appl. Phys. Lett. **55**, 203 (1989)

[23] S. Lanyi, J. Rörök and P. Rehurek, Rev. Sci. Instrum. **65**, 2258 (1994)

# 6 NM LATERAL RESOLUTION IN SCANNING NEAR FIELD OPTICAL MICROSCOPY WITH THE TETRAHEDRAL TIP

J. KOGLIN, U. C. FISCHER AND H. FUCHS

*Westfälische-Wilhelms Universität Münster*

*Physikalisches Institut*

*Wilhelm-Klemm-Str. 10*

*48149 Münster, Germany*

ABSTRACT. We realized a combination of a scanning near field optical (SNOM) and a scanning tunneling microscope (STM) using the tetrahedral tip as a probe. The SNOM and the STM signal are acquired simultaneously during the scan. In the STM mode atomic resolution on pyrolytic graphite is routinely obtained. Simultaneous SNOM/STM investigations of thin silver films evaporated on glass show a lateral resolution of 6 nm in the near field optical signal. Absorption contrast in the optical image is obtained in images of evaporated silver films as well as of patches of purple membrane deposited on indium tin oxide (ITO) as a substrate.

## 1. Introduction

Most experiments in near field microscopy, where the tip was used as a small light source rather than a detector, were made with tapered monomode fibres which were usually coated with aluminum [1]. A small aperture of more than 50 nm in diameter served as small light source. But this kind of illumination is not the only one which can be used in SNOM. Investigations with a small metallic protrusion [2] or complete opaque tips [3,4,5] showed, that these tips are suited as probes for near field microscopy as well.

*M. Nieto-Vesperinas and N. García (eds.), Optics at the Nanometer Scale 247–256.*
© 1996 IBM. Printed in the Netherlands.

We introduced the tetrahedral tip as an efficient apertureless SNOM probe [6]. The body of this tip, as shown in fig. 1, consists of a triangular glass fragment [7]. Due to the fabrication by a successive fracture process the tip and the edges have an unresolved sharpness in the nanometer range. The tip, the adjacent faces (S12, S13 and S23) and two edges (K2 and K3) are coated with a gold film of a thickness of 50 nm. We assume that one edge (K1) remains free or is coated with less metal as expected from the evaporation process [8]. A gold cluster at the apex of the tip seems to serve as an apertureless probe for SNOM and STM which is easily performed simultaneously. We use the tunneling current as a signal to control the distance between tip and sample. The experiments, presented here, show that this tetrahedral tip is suited for near field optical microscopy at ultra high resolution in a combined SNOM/STM mode.

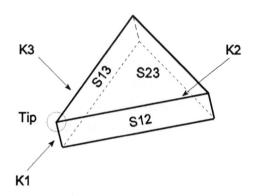

Fig. 1. The tetrahedral tip. The faces S12, S13 and S23, the edges K2 and K3 and the tip are coated with gold. The edge K1 remains free.

## 2. SNOM/STM Setup

The combined SNOM/STM system is based on a commercial beetle type STM [9] and is shown in fig. 2. The tetrahedral tip is mounted onto the z-piezo tube instead of a common STM tip. Light of a laser diode (1 mW, 635 nm) is polarized and focused on a pinhole of 20 $\mu$m. An iris diaphragm limits the aperture and reduces stray light. A fourfold demagnified image of the spot is projected into the tip. The transmitted light is detected by a common Si-photodiode mounted onto a simple commercial light microscope. In the image plane a 200 $\mu$m pinhole masks the tip from stray light. The microscope can be operated in a pure STM mode. A simultaneous SNOM/STM mode proceeds by using the tunnel current for feedback control.

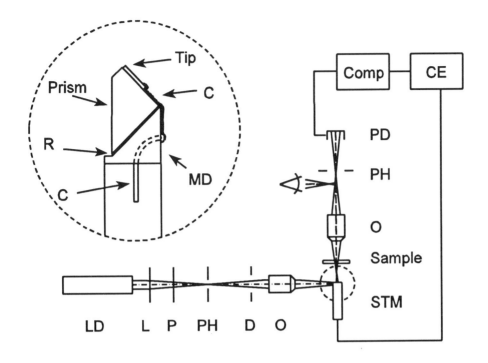

*Fig. 2.* Combined SNOM/STM setup. MD: mounting device, C: connector, R: reflector, LD: laser diode, L: lens, P: polarizer, PH pinhole, D: diaphragm, O: objective, PD: photo diode, CE: control electronic, Comp: Computer

## 3. Results

### 3.1. THE TETRAHEDRAL TIP AS STM PROBE

We tested first the STM mode by imaging highly orientated pyrolytic graphite (HOPG). With the tetrahedral tip the images show atomic resolution routinely. The existence of a gold cluster at the apex of the tip is a basic requirement for the STM mode. The measurements of HOPG and other investigations of metallic objects showed, that a tetrahedral tip used as a STM tip has nearly the same performance as other well prepared Pt/Ir- or W-tips.

250

## 3.2. SILVER FILM ON GLASS

As a test sample for the combined SNOM/STM mode we thermally evaporated a silver film of 50 nm onto a glass substrate at high vacuum ($p = 10^{-6}$ torr). From STM images we found out that the silver grains were typically less than 40 nm wide and less than 8 nm high. Fig. 3 and fig. 4 show a STM and a SNOM image of the silver film which were taken simultaneously in

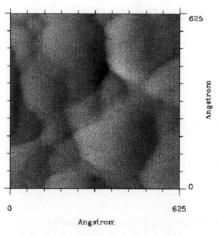

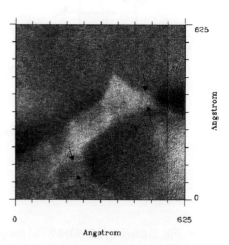

*Fig. 3.* STM image of a silver film of 50 nm evaporated on glass (from [10]).

*Fig. 4.* SNOM image of a silver film. The image was taken simultaneously with the STM image fig. 3 (from [10]).

the combined mode using the tetrahedral tip as probe [10]. In both images characteristic silver grains can be clearly identified. Larger grains show a stronger absorption in the optical signal indicating that the SNOM signal shows a qualitative correlation to the thickness of the silver grains as determined by the STM image. From the absorbance of a 50 nm thick film of silver of 98 % one would expect by simply assuming an exponential absorption law a variation of the SNOM signal of about 37 % for a height difference of the silver film of 5 nm. Therefore the contrast — seen in the SNOM image above — of typically 5 - 10 % is a reasonable value to be due to absorption contrast. Certainly the assumption of an exponential absorption is not adequate in an absorption process induced by a SNOM probe. We cannot explain the absorption as an artefact being due to STM feedback control. We observed that deviations of the distance between tip and sample in the order of few nanometres do not change the light intensity significantly. Approach curves of the tetrahedral tip towards metallic or dielectric samples only showed a slight dependence of the transmitted light intensity on the distance. Insufficiencies of the feedback control can therefore not be a reason for a varying light absorption on the plateaus of silver grains of different thickness.

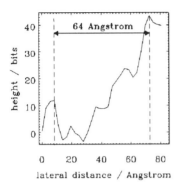
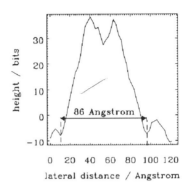

*Fig. 5.* Profile through the lower left part of the SNOM image fig. 4.

*Fig. 6.* Profile through the right part of the SNOM image fig. 4.

Separated features of less than 10 nm in lateral direction are easy to identify at the lower left part and at the right side of the SNOM image. From line profiles as presented in fig. 5 and fig. 6 we determined a point-to-point resolution of 6.4 nm or 8.6 nm respectively. The edge resolution of such SNOM images taken with the tetrahedral tip is less than 1 nm. At the left side of fig. 6 a small feature of around 1 nm at full width at half maximum is shown. We do not base our analysis of resolution on a criterion of single isolated details or calculated edge resolutions. Whether a higher resolution than 6 nm can be obtained will be the object of further investigations.

3.3. SILVER ON INDIUM TIN OXIDE

We investigated silver particles of a silver film of a nominal thickness of 0.5 nm deposited on an indium tin oxide substrate [10]. ITO as a conductive material with a low sheet resistance is thereby suited for STM examinations. Its high transmittance of more than 85% at visible light

252

and its flat surface qualifies ITO at the same time as a substrate used in transmission near field optical microscopy.

In fig. 7 and fig. 8 STM and simultaneously obtained SNOM images of this sample are shown. Like in the SNOM image shown in fig. 4 the resolution of these images is remarkably high. The lateral resolution determined from the distance of separated grain boundaries of the ITO substrate is around 6 nm. The contrast of the apparent ITO grains is inverted compared to the contrast of the image of the silver film on glass (fig. 4). The ITO grains appear bright whereas the boundaries appear dark. We interpret this result by the deposition of the metallic particles at the grain boundaries of the ITO substrate by decoration effects. Consequently the boundaries of the ITO substrate may possess a higher absorption due to the silver particles, and the ITO grains which are free of metal have a higher transmission. On bare ITO we never obtained images showing a comparable contrast.

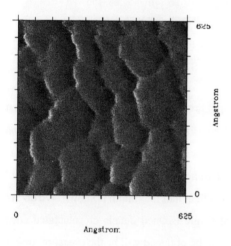 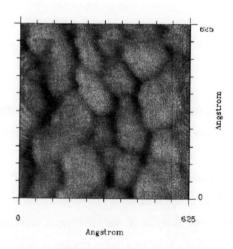

*Fig. 7.* STM image of 0.5 nm silver evaporated on an indium tin oxide substrate (from [10]).

*Fig. 8.* SNOM image. The image was taken simultaneously with the STM image fig. 7 (from [10]).

## 3.4. PURPLE MEMBRANE

Based on these results we started to investigate the purple membrane as a first biological object. The purple membrane is a hexagonal two-dimensional lipoprotein crystal of a periodicity of 6.2 nm. Its chemical composition, its structure and its biological function as a light driven proton pump are known in great detail [11]. The subunit of the purple membrane contains a trimer of bacteriorhodopsin, in which a lysine residue of the membrane protein bacterioopsin forms a Schiff's base with retinal. This retinal-protein complex is the origin of the deep purple colour of this membrane with an absorption maximum at 575 nm at neutral pH shifting to 604 nm at acid conditions. By irradiation the bacteriorhodopsin undergoes a thermo-reversible photochemical cycle with its rate limiting step also depending strongly on environmental conditions. It is a great challenge to near field microscopy to image spectroscopic properties and photochemical processes in bacteriorhodopsin at molecular resolution.

In a first step we studied the dry purple membrane deposited on the ITO substrate. STM images show typical cracks of the membrane with their characteristic angles (see fig. 9) as they are frequently seen also in TEM images [12]. Fig. 10 and fig. 11 show simultaneous STM and SNOM images of the object. A regular structure with a hexagonal pattern at some parts of the sample can be easily identified in the STM image. The periodicity of the structure is not completely resolved — in the most parts of the images the distance of the lattice appears too large. The SNOM images of the purple membrane show an absorption contrast which is definitely due to the thickness of the deposited membrane layers. With increasing thickness of the adsorbate the transmitted light decreases. But the high contrast of the SNOM signal cannot be simply explained by the

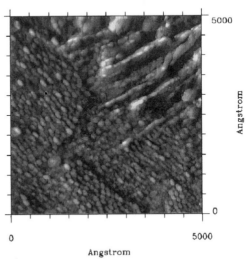

*Fig. 9.* STM image of the purple membrane deposited on indium tin oxide.

absorption properties of the chromophore of the purple membrane, because the density of the retinal amounts to only 0.068 molecules/nm$^2$. This value corresponds to an absorption contrast of about 0.6 ‰ at 575 nm and even less at 635 nm where the images were taken. At the moment the origin of the contrast is unclear. It is conceivable that the difference in refractive index of the ITO substrate and the purple membrane are responsible for the contrast. An investigation using different wavelengths of irradiating light and different conditions of membrane deposition will provide a clue to the origin of the near field optical contrast in these images.

*Fig. 10.* STM image of the purple membrane deposited on indium tin oxide.

*Fig. 11.* Simultaneous SNOM image of the purple membrane deposited on ITO.

## 4. Conclusions

With the tetrahedral tip we are able to perform scanning near field optical microscopy with an apertureless probe. From our examinations of a silver film on glass or silver particles evaporated on an ITO substrate we conclude that the lateral resolution of the tetrahedral tip is around 6 nm which was never reached with aperture-SNOM instruments until now. We determined this resolution value conservatively from the distance of separated features. It appears that the obtained resolution is limited by the size of the metal grains serving as a test object. The tetrahedral tip may imply a better resolution, because details of around 1 nm at full width at half maximum can be imaged.

We obtained a high contrast of the transmitted light signal due to the evaporated silver grains. With increasing thickness of the grains the light intensity decreases. Investigations of the purple membrane showed an absorption contrast which corresponds to the thickness of the deposited membrane patches. Indeed the SNOM signal is correlated to the structures visible in the STM image. The contrast in the SNOM images seems to be too strong to be accounted by the absorption of the chromophore. The origin of the contrast has to be elucidated in further investigation.

With the tetrahedral tip we expect a low thermal load by a local excitation of the sample. Contrary to global illuminations as used in photon scanning tunneling microscopes (PSTM) or other near field optic configurations, a small light source as presented here allows investigations of biological objects which normally are sensitive to high temperatures.

## 5. Acknowledgement

We gratefully acknowledge the support by the Deutsche Forschungsgemeinschaft (German science foundation, Grant Fi-408) and the Carl Zeiss Jena GmbH, Germany. The purple membrane was generously supplied by D. Oesterhelt.

REFERENCES

1. E. Betzig, J.K. Trautmann, T.D. Harris, J.S. Weiner and R.L. Kostelak *Breaking the Diffraction Barrier: Optical Microscopy on a nanometric Scale,* Science 251, 1468-1470 (1991)

2. U. C. Fischer and D. W. Pohl, *Observation of Single-Particle Plasmons by Near-Field Optical Microscopy*, Phys. Rev. Lett. 62 (4), 458-461 (1989)

3. U. C. Fischer, *Resolution and Contrast Generation in Scanning Near Field Optical Microscopy*, in: *Scanning Tunneling Microscopy and Related Methods*, eds.:R. J. Behm et al., 475-461, Kluwer Academic Publishers, Netherlands, 1990

4. M. Specht, J.D. Pedarnig, W.M. Heckl and T. Hänsch, *Scanning Plasmon Near-Field Microscope*, Phys, Rev. Lett. 68 (4), 476-469 (1992)

5. F. Zenhausern, M.P. O'Boyle and H.K. Wickramasinghe, *Apertureless Near-Field Optical Microscope*, Appl. Phys. Lett. 65 (13), 1623-1625 (1994)

6. U. C. Fischer, J. Koglin and H. Fuchs, *The tetrahedral Tip as a Probe for Scanning Near-Field Optical Microscopy at 30 nm Resolution*, J. Microscopy 176, Pt 3, 231-237, (1994)

7. J. Koglin, U.C. Fischer, K.D. Brzoska, W. Göhde and H. Fuchs, *The Tetrahedral Tip as a Probe for Scanning Near-Field Optical Microscopy and for Scanning Tunneling Microscopy*, in: *Photons and Local Probes*, eds.: O. Marti and R. Möller, 79-92, Kluwer Academic Publishers, Netherlands, 1995

8. H. U. Danzebrink and U. C. Fischer, *The Concept of an Optoelectronic Probe for Near Field Microscopy*, in: *Near Field Optics*, eds.: D. W. Pohl and D. Courjon, 303-308, Kluwer Academic Publishers, Netherlands, 1993

9. K. Besocke, *An easily operable Scanning Tunneling Microscope*, Surf. Sci. 181, 145 (1987)

10. J. Koglin, U.C. Fischer and H. Fuchs, *Scanning Near Field Optical Microscopy with the Tetrahedral Tip az a Resolution of 6 nm*, Journal of Biomedical Optics 1(1), (1996), in print

11. D. Oesterhelt, J. Tittor and E. Bamberg, *A Unifying Concept for Ion Translocation by Retinal Proteins*, Journal of Bioenergetics and Biomembranes, Vol. 24, No. 2, 181-191, (1992)

12. D.-Ch. Neugebauer and H. P. Zingsheim, *The two Faces of the Purple Membrane*, J. Mol. Biol. (1978) 123, 235-246

# AN APERTURE-TYPE REFLECTION-MODE SNOM

C. Durkan & I.V. Shvets
Department of Physics,
Trinity College,
Dublin 2,
Ireland

A reflection-mode aperture SNOM based on external collection of the reflected light is presented. The light detection is based on an elliptical mirror set-up, with the tip and sample at one focus, and a PMT at the other. Tapered metal-coated fibres are used as the SNOM probes. These are produced using an in-house computer-controlled pulling machine. We operate the SNOM at a wavelength of 685 nm. Shear-force distance regulation is used. We have developed a distance regulation system whereby a dither amplitude of the fibre tip of 5 nm can easily be detected with a signal/noise ratio of better than 10. Images of a range of samples, including a cross diffraction grating and an aluminium pattern on glass indicate a shear-force resolution in the range 20-30 nm, and an optical resolution better than 50 nm.

## 1. Introduction

A reflection-mode SNOM has been developed for the imaging of opaque samples. This incorporates a shear-force[1,2] based distance regulation system for the imaging of conducting as well as nonconducting samples. Many reflection-mode SNOMs today use the same uncoated fibre tip for both illumination and collection[3]. We decided to take the alternative route of collecting the reflected light[4,5] from the sample by means of an elliptical mirror which is coaxial with the fibre tip. The advantages of such a setup are (i) for applications where a well defined small spot of light is required- for eg., in any local surface modification experiments where one wants to optically or thermally "write" features on a surface, and (ii) transmission mode SNOM may be simultaneously done.

## 2. Experimental, results and discussion

The experimental setup is as shown in figure 1. The fibre tips, which are produced with an in-house computer-controlled pulling machine[6] are coated with 100-150nm

257

*M. Nieto-Vesperinas and N. García (eds.), Optics at the Nanometer Scale 257–261.*

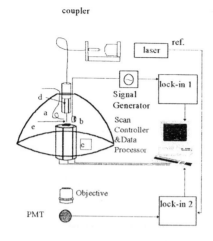

coupler

laser    ref.

lock-in 1

Signal
Generator

d
a
e    b
c

Scan
Controller
&Data
Processor

Objective

PMT

lock-in 2

*Figure 1*

of aluminium. The method of the shear-force detection is based on using the fibre tip(e) as a cylindrical lens to deflect and focus a beam from a second optical fibre(a), onto the edge of an aperture(b) in front of a photodiode as shown in figure 1. The tips are oscillated laterally by a dither piezo(d) by approximately 5nm peak-peak at a frequency just below mechanical resonance, which is typically 10-20kHz. This is easily detected with a lock-in amplifier, with a signal-to-noise ratio of 10-15. As the working distance for shear-force is of the order 5-10nm this means the vertical resolution of the detection scheme is a few angstroms. Figure 2 shows shear-force approach curves for different fibre oscillation

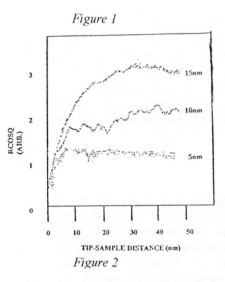

RCOSQ
(ARB.)

3

2

1

0

15nm

10nm

5nm

0    10    20    30    40    50

TIP-SAMPLE DISTANCE (nm)

*Figure 2*

*Figure 3*

amplitudes. Figure 3 (scan size=1.2μm$^2$) shows a typical shear-force image of a grating structure. This consists of a carbon cross grating replica with a series of v-shaped grooves of period 460nm, groove depth 40-80nm and groove width 80nm. Analysis of line profiles indicates a lateral resolution of better than 30nm.

The optical signal for SNOM is collected with an elliptical mirror which is coaxial with the SNOM tip. This enables the collection of a large amount of the reflected light, in a symmetric manner. This collection symmetry helps to reduce shadowing effects. This arrangement has been found to be superior to the situation where a single objective collects the reflected light[7]. For SNOM, the typical detected optical

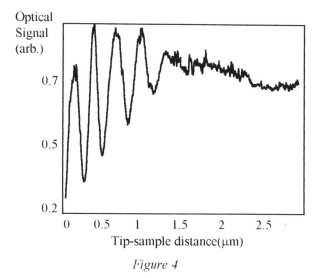

Optical Signal (arb.)

0.7

0.5

0.2

0    0.5    1    1.5    2    2.5

Tip-sample distance(μm)

*Figure 4*

signal level on such samples is <1nW. This corresponds to a transmission through the tips of the order $10^{-6}$. On monitoring this signal while the tip is being sample, interference oscillations are clearly seen, with the last maximum at a tip-sample separation of $\lambda/4$ as shown in figure 4. This is due to interference between

light emmitted from the tip and light reflected from the sample.

Figure 5 shows the shear-force and SNOM images of a portion of this sample (scan size=870*930nm). Figure 5a shows the shear-force image, where the grating lines are clearly resolved, and figure 5b shows the SNOM image of the same area. In some regions, the grating lines may be seen, as well as a series of bright dots, corresponding to depressions in the topography. The size of these features is 40nm, and they are seperated by 60nm. These type of features were also seen on the sample by STM(Scanning Tunneling Microscopy) and TEM(Transmission Electron Microscopy). Therefore, from the SNOM image, an optical resolution of better than 40nm may be inferred. On inspection of the force image, it is clear that the grating lines appear raised (in the gray-scale image, bright corresponds to raised features, and dark to depressed features) instead of depressed as they actually are. This is simply a tip shape effect-when the tip is descending into a groove, and ascending back out during scanning, it experiences an influence (shear-force damping) primarily due to one side of the groove. When the tip is centered on the groove, however, it now experiences a strong influence from both sides, so it will withdraw, in order to maintain a constant oscillation amplitude. Depending on the shape of the tip and the feedback characteristics, the tip may withdraw enough to produce a contrast reversal. Figure 6 shows two shear force images taken of the sample with the same tip (scan size=1.5μm$^2$). The image in figure 6a was taken first, followed by the image in figure 6b. In between the taking of these images, there was a tip-crash, during which the tip was reformed, and apparently sharpened. The scan time for both images was the same, so inadequecies in the feedback loop may be ruled out as primarily causing the contrast reversal.

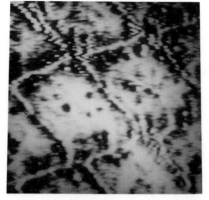

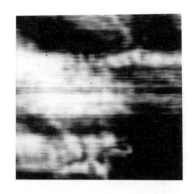

*Figure 5a*                     *Figure 5b*

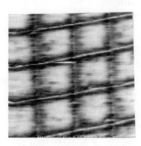

*Figure 6a*                     *Figure 6b*

Since in the grating sample all the optical contrast arises due to the topography of the surface, we decided to carry out experiments on a sample containing optical contrast as well as topography. The sample was produced by placing a droplet of water containing a suspension of $0.5\mu m$ diameter latex spheres onto a glass substrate, and evaporating aluminium onto this. Subsequent removal of the spheres leaves a well defined triangular pattern of al on glass. Figure 7 shows a SNOM image (scan size=$1.2\mu m^2$) of a portion of this sample. In this case, the thickness of the al film was 40nm. The pattern left behind by the spheres is clearly visible.

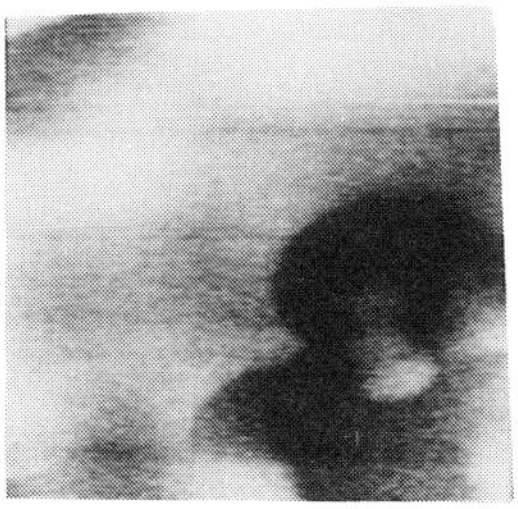

### 3. Conclusions

*Figure 7*

A reflection-mode SNOM has been presented, with mirror collection of the reflected light, demonstrating optical resolution in the range 40nm, on samples both with and without optical contrast. This is comparable to that found in transmission-mode systems. Measurements on the distance dependance of the shear-force interaction have been presented, as well as images showing a shear-force resolution of 20-30nm.

### 4. References

1. Betzig, E., Finn, P.L. and Weiner, J.S. (1994) Appl. Phys. Lett. **60**, 2484

2. Durkan, C., Shvets, I.V. (1996) J. Appl. Phys, *in press.*

3. Bielefeldt, H., Horsch, I., Krausch, G., Mlynek, J., Marti, O. (1994) Appl. Phys. A**59**, 103

4. Grober, R.D., Harris, T.D., Trautman, J.K., Betzig, E. (1994) Rev. Sci. Instrum. **65**(3), 626

5. Fischer, U.Ch., U.T. Durig and D.W. Pohl (1988) Appl. Phys. Lett. **52** 249

6. Shvets, I.V., Madsen, S. (1995) "Computer-controlled fibre-pulling machine for fabrication of optical-fibre probes for Near-Field Microscopy", Irish Patent No. S63296

7. Durkan, C., Shvets, I.V. (1995) in *Photons & Local Probes*, NATO ASI Series, ed. Kluwer.

# SURFACE MODIFICATIONS VIA PHOTO-CHEMISTRY IN A REFLECTION SCANNING NEAR-FIELD OPTICAL MICROSCOPE

STEEN MADSEN, TOM OLESEN and JØRN M. HVAM*

*Danish Micro Engineering A/S, DK-2730 Herlev, Denmark*

*\*Mikroelektronik Centret, DTU, DK-2800 Lyngby, Denmark*

## ABSTRACT

A reflection scanning near-field optical microscope (SNOM) utilizing counter-directional light propagation has been used to modify the surface topography via photo-chemical processes. The effect of photo induced surface modifications was demonstrated on a 9-chloroanthracene crystal surface. The crystal was illuminated through an uncoated tapered fibre probe in the reflection SNOM. A shear-force microscope was used to control the probe-sample distance during the writing process, and to simultaneously image the topographical and optical properties with sub-wavelength resolution.

263

M. Nieto-Vesperinas and N. García (eds.), Optics at the Nanometer Scale 263–275.
© 1996 IBM. Printed in the Netherlands.

# 1. Introduction

The invention of the scanning tunneling microscope (STM) [1] and the atomic force microscope (AFM) [2] started a new era of high resolution microscopy. The STM and the AFM belong to the group scanning probe microscopes (SPM). Another member of the SPM family is the scanning near-field optical microscope (SNOM) [3]. The image formation in conventional optical microscopy is based on propagating electromagnetic fields, whereby the resolving power is set by the Abbe diffraction limit. Therefore, in general, one can not resolve features smaller than 0.2 μm using even the most advanced conventional optical microscope. In the SNOM, the imaging is based on the evanescent field mode, whereby the achievable lateral resolution surpasses the far-field diffraction limit. A lateral resolution of 20 nm, photoluminescence spectroscopy and sub-micron surface modifications has successfully been demonstrated using SNOM.

We have built a reflection SNOM where the sample is illuminated through an uncoated tapered fibre probe. As well as acting as a subwavelength size light source, the probe is also used as the detector. In this way both transparent and opaque samples can be analysed. Compared to the more frequently used reflection aperture SNOM using a metal coated probe, and an external detection system, our configuration completely avoids shadowing effects from the fibre probe itself.

# 2. Experimental technique

The experimental setup is shown in figure 1. The setup is based on the DME - Rasterscope SNOM [4]. A main component of the instrument is the SNOM probe, which is an uncoated tapered single-mode optical fibre. The probes are drawn in a commercially available Micropipette Puller [5].

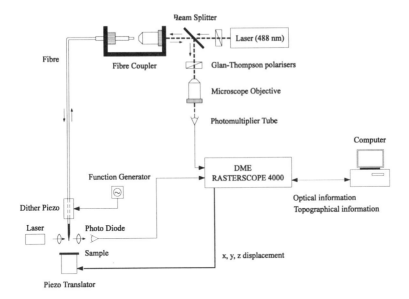

*Figure 1.* Experimental setup for the reflection scanning near-field optical microscope with shear-force distance regulation. A cross polarisation detection system is implemented to increase the signal to noise ratio of the near-field detection system.

To characterise the size and shape of the probes, a scanning electron microscope (SEM) was used. An SEM micrograph of a SNOM fibre probe is seen in figure 2.

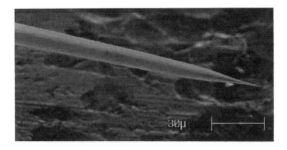

*Figure 2.* SEM micrograph of an uncoated fibre probe. The length of the inserted bar is 30 μm. The radius of curvature of the very end is typical 10-15 nm.

The light emitting properties of each individual probe was studied by looking at the far-field diffraction pattern. This pattern consists of concentric rings around the fibre axis. For a sharp probe, the intensity of the centre spot was low, whereas in the case of a more blunt probe, the intensity of the centre spot was much higher. The probes were laterally dithered near their mechanical resonance frequency, and the rectified AC part of the shear-force signal was used as a feed-back signal to keep the probe-sample distance constant while scanning. In this way topographical and near-field optical images are recorded simultaneously.

The purpose of these preliminary experiments was to demonstrate how a SNOM can be used to modify the surface topography via photo-chemistry. The photodimerization of 9-chloroanthracene was chosen for this first demonstration. The mechanism behind photo-dimerization is absorption in the crystal. A 9-chloroanthracene crystal has a pronounced absorption peak around 500 nm [6]. For this reason a stabilized air cooled $Ar^+$ laser ($\lambda$=488 nm) was chosen as the light source. Linearly polarised light is coupled into the fibre through a 50:50 beam splitter. Most of the incident light is coupled out, and thereby lost, through the sides of the tapered probe. However, a small part scatters from the very end of the probe and is reflected back into the fibre. The backward propagating light is directed towards a photomultiplier tube via the same beam splitter arrangement. Intensities in the forward and backward propagation directions are typically 1.6 mW and 0.2 μW, respectively. A cross polarisation detection system is implemented to minimize the influence of unwanted reflections from coupling lenses, fibre input interface etc.. [7] This detection technique relies on the fact that a substantial part of the incoming linearly polarised light is coupled to the orthogonal component while propagating twice through the tapered fibre probe. Our system is implemented with two Glan-Thompson linear polarisers having their optical axis oriented perpendicular to each other.

## 3. Results

The lateral resolution of the reflection SNOM was tested on a cross grating structure. The period of the grating is 200 nm. Figure 3 shows a 2x2 $\mu m^2$ scan area, where (a) is the topographical image, and (b) is the near-field optical image. The height variation in (a) is 80 nm. Bright areas in the near-field optical image correspond to high reflectivity from the probe-sample interaction region. The diameter of the bright areas in (b), and the distance between them, is approximately half a period, i.e. ~100 nm. This yields an optical resolution of at least 100 nm.

a)                                        b)

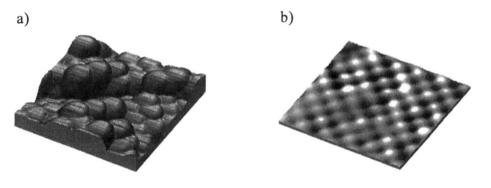

*Figure 3.* Images of a cross grating. The period of the grating is 200 nm. The scan area is 2x2 $\mu m^2$. Image (a) shows topography, (b) is the corresponding near-field optical image. The height variation in (a) is 80 nm.

The optical contrast seen in 3b is, at least in part, due to the topography of the sample. Decoupling topographical information from optical information is a general problem in near-field microscopy. In practice, this coupling can not be completely avoided, but restricting the scanning to relative flat samples will reduce the problem. As a consequence, the ideal sample for testing a near-field microscope is a topographically completely flat sample surface exhibiting only optical contrasts.

A single crystal of 9-chloroanthracene was now mounted in the SNOM. A typical set of images is shown in figure 4, where 4a is the topography and 4b is the near-field optical image.

a)                                             b)

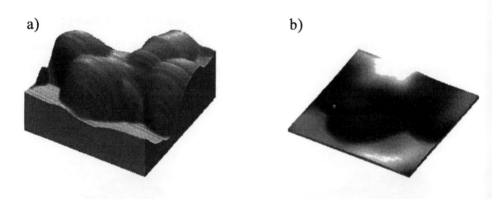

*Figure 4.* Images of a 9-chloroanthracene crystal surface. The scan area is 1x1 μm². Image (a) is the topography, (b) is the corresponding near-field optical image.

The total height of the topographical image is 121 nm. The size of the surface structures was measured to be 250-500 nm. During the imaging, expected changes in the surface topography were observed. This is illustrated in figure 5. The four topographical images are all images of the same surface area, recorded with a 3 minute interval. While imaging the surface topography, the sample was illuminated through the uncoated fibre probe ($\lambda$ = 488 nm). The intensity coupled into the fibre probe was 1.2 mW. The photo-chemically induced surface modifications could easily be stopped simply by switching off the illumination.

A cross-sectional profile of the upper left image (T = 0 min.) and the lower right image (T = 9 min.) is shown in figure 6. The profile is taken from the lower left to the upper right corner in both figures. Comparing the two profiles, represented by the dotted and the solid line, the change in surface topography is evident. The zero point on the z scale (height) has been set automatically by the microscope for optimum performance.

The average height of the four images in figure 5 have been calculated, and the result is plotted in figure 7. The photodimerization of the crystal is seen to cause an increase in the average height of the surface topography. A second order polynomial regression has been fitted to the experimental data.

0 min.        3 min.

6 min.        9 min.

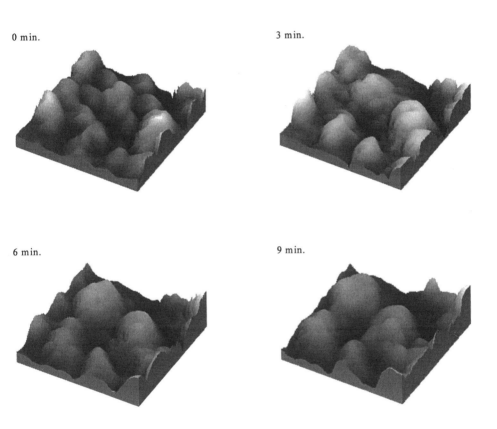

*Figure 5.* Four topographical images of the same area on a 9-chloroanthracene crystal surface. The scan area is 5x5 μm². The images were recorded with a 3 minute interval.

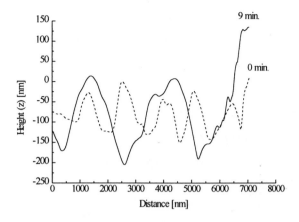

*Figure 6.* Cross-sectional profiles of two of the images in figure 5, T = 0 min. (dotted line) and T = 9 min. (solid line). The point z=0 has been set automatically for optimum performance in the piezo-electric scanner tube.

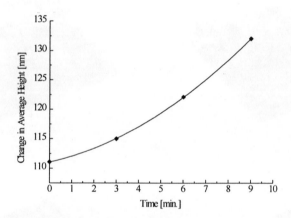

*Figure 7.* Change in average height vs. time calculated from the four images in figure 5. A second order polynomial regression has been fitted to the four experimental data.

Instead of scanning the crystal, and thereby expose a relatively big area, the fibre probe was now fixed at a certain position, approximately 10 nm above the sample surface. The sample was exposed at this position in 20 min., and subsequently scanned without light coupled into the probe.

a)

b)

c)

d)

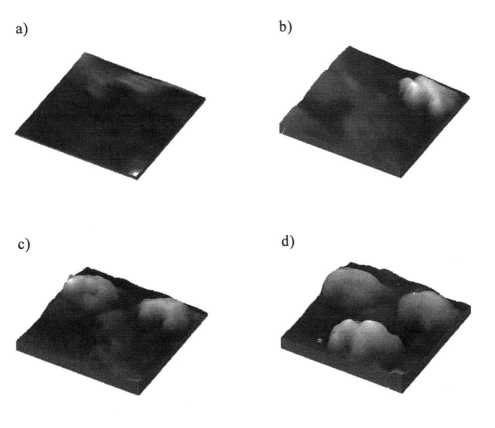

*Figure 8.* Four topographical images of the same surface area. The scan area is 5x5 μm². (a) is the unmodified surface, (b) shows the first grown feature, (c) two and (d) three grown features. coupled into the probe.

The probe was then moved to two other positions where the procedure was repeated. The result is shown in figure 8, where (a) is the unmodified surface, (b) shows the first grown feature, (c) shows two and (d) shows three grown features. The scan area in all four images is 5x5 μm². The average height of the first feature is 91 nm. This point has been added as a fifth point to the four points of figure 7, and the result is plotted in figure 9. The first point in figure 7 (T=0 min.) has been assigned the value Z=0. Again, a second order polynomial regression has been fitted to the experimental data.

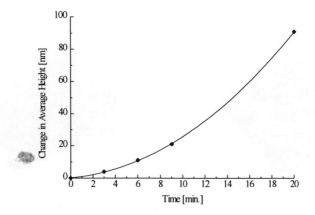

*Figure 9.* Change in average height vs. time. The first four points (T=0 min., T=3 min., T=6 min. and T=9 min.) represent the change in average height calculated from figure 5. The fifth point (T=20 min.) is the average height of the first grown features, see figure 9b. A second order polynomial regression has been fitted to the five experimental data.

The lateral dimensions of the grown structures vary along different directions. Along x (left to right) the size is ~2.5 µm, whereas along y the size is between 700 and 1000 nm. Since the emitted light from the probes was symmetric around the fibre axis, this non-symmetric shape could be due to the material properties of the 9-chloroanthracene. The relatively large dimensions of the grown features can be explained from the fact, that an uncoated fibre probe was used to illuminate the crystal. Assuming that the intensity emitted from a coated fibre probe is sufficient to cause photodimerization, such a probe could perhaps have reduced the dimensions of the features.

After growing the three positive features, 16 images were recorded over the following 31 minutes. While scanning the crystal was illuminated through the fibre probe, and the change in surface topography was studied. Image number 1 and 16 are shown in figure 10a and 10b, respectively. The positive features are clearly seen in both images, but due to surface modifications the edges appear more blunt in 10b. The smaller background features in 10a are also strongly modified and appear much bigger in 10b.

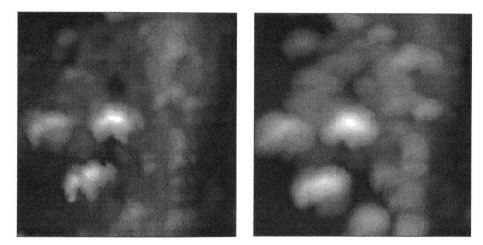

*Figure 10.* Two topographical images of the modified crystal surface recorded with approx. 30 minute interval. The scan area is 10x10 $\mu m^2$.

## 4. Conclusions

A reflection SNOM has been used to demonstrate photo-chemistry on a single crystal of 9-chloroanthracene. The surface was illuminated through an uncoated fibre probe, whereby positive features were locally grown at the crystal surface. The lateral dimensions vary between 0.7 and 2.5 $\mu m$. Using a metal coated fibre probe with a smaller aperture could possibly reduce the size of the features because of a more confined fibre mode. Photochemistry, which can be considered as a special kind of optical lithography, has successfully been used to write spots with a diameter of 140 nm in an azobenzene LB film. The processed region showed a change in absorption of 8% [8].This leads directly to an area of huge interest, namely high density data storage, where also photo-chemistry is expected to play an important role.

274

**Acknowledgement**

This work was supported by the Danish Academy of Technical Sciences. The authors would like to thank Prof. Dr. Gerd Kaupp, University of Oldenburg, for valuable discussions and suggestions, and for providing the samples.

# References

1. Binnig, G., Rohrer H., Gerber Ch. and Weibel E. (1982) Surface Studies By Scanning Tunneling Microscopy, Phys. Rev. Lett. **49**, 57-61

2. Binnig, G., Quate C.F. and Gerber Ch. (1986) Atomic Force Microscopy, Phys. Rev. Lett. **56**, 930-933

3. Pohl D.W., Denk W. and Lanz M. (1984) Optical stethoscopy: Image recording with resolution $\lambda/20$, Appl. Phys. Lett. **44**, 651-653

4. DME - Rasterscope SNOM, Danish Micro Engineering A/S, DK-2730 Herlev, Denmark

5. Model P-2000 Laser Based Micropipette Puller, Sutter Instrument Company, CA, USA

6. Prof. Dr. Kaupp G. (Private communication)

7. Bozhevolnyi S.I., Keller O. and Xiao M. (1993) Control of the tip-surface distance in near-field microscopy, Appl. Opt. **32**, 4864-4868

8. Jiang S., Ichihashi J., Monobe M., Fujihira M. and Ohtsu M. (1993) Presented at NFO II

# NEAR-FIELD DIFFRACTION MICROSCOPY WITH A COHERENT LOW-ENERGY e-BEAM: FRESNEL PROJECTION MICROSCOPE

VU THIEN BINH[a], V. SEMET[a], N. GARCIA[b] and L. BITAR[a,b]
a) *Laboratoire d'Émission Électronique, DPM - URA CNRS Université Claude Bernard Lyon 1. 69622. Villeurbanne. France*
b) *Fisica de Sistemas Pequenos, CSIC Universidad Autonoma de Madrid, CIII, 28049 Madrid Spain*

*The specific characteristics of e-beams field emitted from nanotips, which are atom-sources of electrons, are fully exploited in a versatile low-energy electron projection microscope: the Fresnel projection microscope. Observations of nanometric fibres of carbon and of organic materials (polymers and RNA), with working voltages around 200 V, show details less than one nanometer without any magnetic shielding. The experimental results are interpreted within the framework of electron optics analysis and by taking into account the field emission properties of nanotips.*

## 1. Introduction

The aim of this article is to outline the physics of the image formation in a Fresnel projection microscope (FPM) [1] which uses a nanotip [2,3] as a coherent electron source at ~200 V. The different aspects of image formation will be discussed in order to explain why FPM experimental observations of nano-objects such as fibres of carbon [1], polymer [4] and RNA [5] can be achieved in the direct space, with resolving details smaller than 1 nm.

The working voltage for the FPM in the range of 200 V means that the electrons incident on the object have a mean free path less than 1 nm [6] and thus, to the first order, objects in the range of 2 to 10 nm will be opaque [4]. As well the de Broglie wavelength ( $\lambda$ (nm) $\approx 1.226 / V^{1/2} = 2\pi / k$ ($k$ is the wave vector) ) is still less than 0.1 nm so that multiple diffraction effects will not obscure all details in the nanometer range. These points together indicate that a basic understanding of the experimental images and its resolution in the FPM can be found by considering the images as due to the Fraunhofer and Fresnel diffractions from a point source, by a mask projected on a screen [7]. Intrinsic properties in both cases will be pointed out in order to show their range of validity. A discussion on the notion of holography within the field of projection microscopy is also included.

*M. Nieto-Vesperinas and N. García (eds.), Optics at the Nanometer Scale 277–296.*

## 2. The Fresnel Projection Microscope

The Fresnel projection microscope is the result of the combination of a nanotip, which is an atom-source of electrons with specific properties, and a projection microscope. The projection or shadow microscope [8] is essentially a lensless microscope based on the radial propagation of an e-beam from a point source (Fig. 1). The projection image has a magnification factor M given by:

$$M = \frac{i}{o} \approx \frac{D}{d},\qquad(1)$$

where i and o are the image and object dimensions and D and d are the distances of the projection point to the screen and to the object respectively. Eq. (1) shows that the magnification increases by approaching the object to the projection point and it can reach values in the range of $10^7$-$10^6$ for projection point-object distances between 10 nm to 100 nm, with the screen located ~10 cm away. With the recent technological developments due to scanning tunneling microscopy (STM), tip-sample distances of less than 1 nm can now be routinely handled by using piezodrives for controlled nanometric displacements [9].

The nanotips are protusions grown on top of microscopic tips and ending in one atom. In concomitance with the nanometric size of the protrusions, the field emitting area which is the last atom of the protrusion leads to specific properties for the field emitted e-beam [3,10]. Among them,

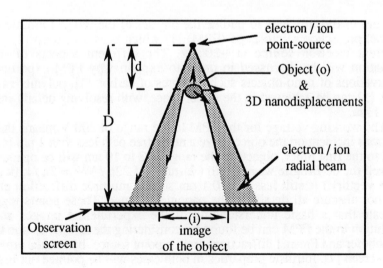

*Figure 1.* Schematic drawing of field emission projection microscope

those of interest for projection microscopy are:

- The size of the emitting area is limited to the last apex atom of the protrusion and the position of the virtual source, which is the projection point, is within nanometric distance from the actual surface.

- The total beam opening in the vicinity of the emitting atom is less than ~20° and it is self-focused to ~5° at macroscopic distance away from the apex due to the focusing lens effect of the whole tip.

- The field emission current is stable to within 0.25% for duration from several hours up to ~10 hours inside ultra-high vacuum.

- The energy spread is ~100 meV at room temperature due to field emission process from localised bands

From the practical point of view, the Fresnel projection microscope is composed essentially of the nanotip, with its surroundings (heating, cooling and temperature measurement), in front of an object holder system which consists of piezo-drive ceramic-based motors allowing a 3D movement in the centimetre range with sub-nanometric steps. This mechanical nano-displacement system makes possible the choice of the imaging zone over a large area by the x,y displacements and a continuous zooming on a chosen area by approaching the object until nanometric distances to the nanotip. Controlled bending of the ceramic piezo-tubes gives the absolute calibration for the x,y scales at the object plane. Note that these scales are determined without any reference to the source-object distance, which is difficult to determine. This means that the scales given in the FPM experimental images below are independent of the imaging process and are related only to the piezo-ceramics calibration accuracy [9]. The image is projected on a fluorescent screen, equipped with a micro-channel plate placed ~10 cm from the nanotip. The projected images are recorded by a video camera with image processing on-line. The volume of the whole effective system - including nanotip, object holder and visualisation screen - is around 1 dm$^3$ and it is inside a ultra-high vacuum chamber. The whole apparatus is isolated from the external vibrations with a pneumatic system. There is no magnetic shielding incorporated in the microscope chamber.

## 3. Experimental observations

The experimental observations were performed with three categories of samples: carbon [1], synthetic polymer [4] and biological RNA fibres [5]. The common feature of these samples is that they are all standing alone objects with a diameter in the range of 1 to a few nanometers.

Carbon fibres were observed directly from commercial holey-carbon film deposited on 3 mm TEM grids. This sample structure allows not only the observation of fibres, but also of holes in the range from μm down to nm. For synthetic polymer and biological samples, the specimen preparation for FPM observation is simply a two-step process: (1) dissolution of the samples in a solvent, then (2) deposition of a drop of this solution on a holey-carbon gold grid. After evaporation of the solvent, the probability of

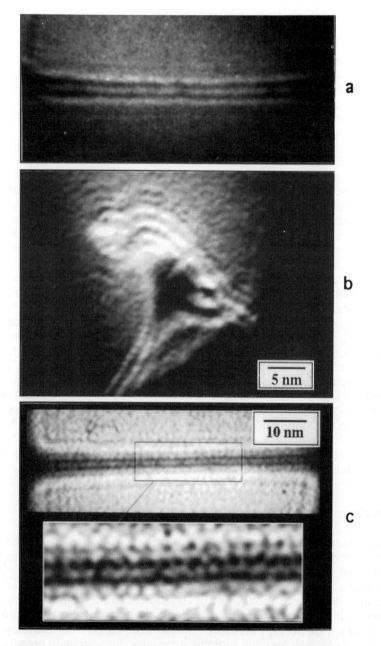

*Figure 2.* Fresnel projection microscopy images of standing alone fibres:
a) carbon single fibre of ~1.5 nm diameter, b) synthetic polymers [PS-95%)- (PVP-5%)
forming a nano-clew and c) biological macromolecules [poly A-poly U - RNA]

having macromolecules stretching across a hole is rather high, allowing easy localisation and observation by FPM of standing alone fibres. Note that no other specimen preparation, such as staining or metal coating for example is necessary,

Figure 2 shows some examples of the FPM images of these objects.

## 4. Fresnel and Fraunhofer diffractions

Introduced by Francesco Maria Grimaldi (1618-1663), diffraction is associated with the obstruction of a single wave, by either a transparent or opaque object, which result in cast shadows differing from the sizes predicted by geometrical optics. The distinction with interference is that this latter may be associated with the intentional formation of two or more light waves that are analysed in their region of overlap.

The analytical treatment of the diffraction theory was developed by Kirchhoff in 1887 and it can be reformulated into a vector theory. Referring

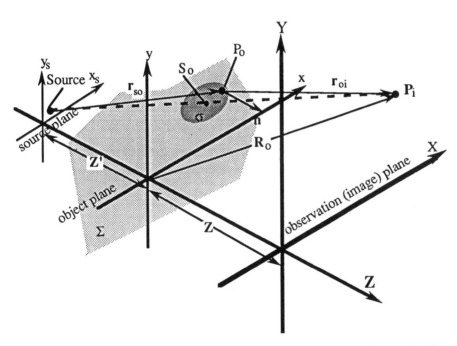

*Figure 3.* Geometry for the Fresnel-Kirchhoff diffraction formula. $(x_s, y_s)$, $(x, y)$ and $(X, Y)$ are the source, object and observation planes respectively

to Fig. 3, given an incident wave $\varphi$ on the object plane, the optical wave at a point $P_i$ located at $R_0$ can be calculated in terms of the wave's value on a surface S constructed around the observation point $P_i$. To obtain a solution to the wave equation at the point $P_i$ we select as a Green's function a spherical wave with amplitude 1, denoted $\Psi$, expanding about the point $P_i$. At the point $P_0$, the Green's function is given by

$$\Psi(P_0) = \frac{e^{-ik \cdot r_{oi}}}{r_{oi}}. \tag{1}$$

Within a volume V surrounding $P_i$ but excluding $P_i$, the Green's function $\Psi$ and the incident wave $\varphi$ must satisfy the scalar Helmholtz propagation wave equation in free space, thus:

$$\iiint_V (\Psi \nabla^2 \varphi - \varphi \nabla^2 \Psi)\, dv = \iiint_V (\Psi \varphi k^2 - \varphi \Psi k^2)\, dv. \tag{2}$$

The right side of (2) is identical equal to zero and thus (2) can be rewritten, by using the Green's theorem, into

$$\iint_S (\Psi \frac{\partial \varphi}{\partial n} - \varphi \frac{\partial \Psi}{\partial n})\, ds = 0. \tag{3}$$

By evaluating the Green's function and its derivative, (3) becomes

$$\varphi(P_i) = \frac{1}{4\pi} \iint_S \left[ \frac{e^{-ik \cdot r_{oi}}}{r_{oi}} \frac{\partial \varphi}{\partial n} - \varphi \frac{\partial}{\partial n} \left( \frac{e^{-ik \cdot r_{oi}}}{r_{oi}} \right) \right] ds. \tag{4}$$

Relation (4) is called the integral theorem of Kirchhoff. The integrand is greatly simplified within the Kirchhoff boundary assumptions about the incident wave $\varphi$ and its derivative $\partial \varphi / \partial n$ :
- inside the aperture $\Sigma$ of the object, $\varphi$ and $\partial \varphi / \partial n$ have the values they would have if the object screen is not in place,
- on the portion of the object not in the aperture, $\varphi$ and $\partial \varphi / \partial n$ are identically zero.
These two boundary conditions imply that the field is zero everywhere directly behind the object except in aperture $\Sigma$, that makes $\Psi$ and $\partial \Psi / \partial n$ discontinuous on the boundary of the aperture. Substituting them into (4) yields

$$\varphi(P_i) = \frac{iA}{\lambda} \iint_\Sigma \frac{e^{-ik \cdot (r_{oi} + r_{so})}}{r_{oi} r_{so}} \left( \frac{\cos(n, r_{oi}) - \cos(n, r_{so})}{2} \right) ds. \tag{5}$$

This integral relation over $\Sigma$ is called the Fresnel-Kirchhoff diffraction formula or Huygens-Fresnel integral. A physical understanding can be obtained by rewriting (5) as

$$\phi\,(P_i) = \iint_\Sigma \Phi(r_{so})\,\frac{e^{-ik\cdot r_{oi}}}{r_{oi}}\,ds\;, \qquad (6)$$

$$\Phi(r_{so}) = \frac{i}{\lambda}\left[A\,\frac{e^{-ik\cdot r_{so}}}{r_{so}}\right]\left(\frac{\cos(n,r_{oi}) - \cos(n,r_{so})}{2}\right). \qquad (7)$$

Thus, (6) means that the field at $P_i$ is due to the sum of secondary Huygens' sources in the apertures of $\Sigma$. They are point sources radiating spherical waves of the form

$$\Phi(r_{so})\,\frac{e^{-ik\cdot r_{oi}}}{r_{oi}}\;,$$

with amplitude $\Phi(r_{so})$ defined in (7) in which the imaginary constant means that they are phase shifted by $\pi/2$ with respect to $\phi(r_{so})$, the incident wave having an amplitude A,

$$\phi(r_{so}) = A\,\frac{e^{-ik\cdot r_{so}}}{r_{so}}\;.$$

The obliquity factor

$$\frac{\cos(n,r_{oi}) - \cos(n,r_{so})}{2}\;,$$

causes the secondary sources to have only a forward-directed radiation pattern.

Two approximations can be made to obtain analytic expressions of (5). These lead to Fraunhofer and Fresnel diffractions which allow discussion of the general properties of image formation in the projection microscope and to find the departure of transmitted wave from the geometrical shadow image given by the optical path. In the projection microscope configuration the distance between the projection source and the detection screen is fixed at macroscopic distances, in the range of centimetres. This implies that the differentiation between Fraunhofer and Fresnel diffraction will be determined only by the distances between the projection source and the object. In Fraunhofer diffraction we require that both the source and the observation point be far from the object. Note that the distance scale is relative to the dimensions of the object, i.e. the overall dimensions of the object is very much smaller than the distances to source or observation point, so that the incident waves can be approximated by plane waves. Under Fresnel diffraction the source is close to the object and the observation point is far away.

## 4.1. FRAUNHOFER DIFFRACTION

Under Fraunhofer diffraction the wave incident on the object $\Sigma$ is a plane wave. A consequence of this requirement is that, at each observation point, the entire wavefront passing through the object contributes to the observed diffraction. For the diffraction integral (5) to have a finite non-zero value the requirement is that the phase changes are small over the whole object,

$$k\left| \mathbf{R}_0 - \mathbf{r}_{oi} \right| \ll k\,\mathbf{R}_0 \; .$$

This approximation can be seen to be equivalent to assuming that the object is small compared to the projection distance $R_0$. Within this approximation the diffraction integral (5) becomes

$$\varphi\,(P_i) = \frac{iA}{\lambda}\,\frac{e^{-i\mathbf{k}\cdot\mathbf{R}_0}}{R_0}$$

$$\times \iint_\Sigma f(x,y)\,\exp\left[+ik\left(\frac{xX+yY}{R_0} - \frac{x^2+y^2}{2\,R_0}\right)\right]$$

$$\times \left(\frac{\cos(\mathbf{n},\,\mathbf{r}_{oi}) - \cos(\mathbf{n},\,\mathbf{r}_{so})}{2}\right)\,dxdy, \tag{8}$$

where $(x,y)$ are the co-ordinates at the object plane and $(X,Y)$ those at the observation plane. The complex transmission function $f(x,y)$ of the aperture is introduced here to allow a very general aperture to be treated, it would be a real function if it describes a variation in adsorption of the aperture for example.

As the observation point is far from the screen, mathematically the second term in the exponential can be neglected

$$k\,\frac{x^2+y^2}{2\,R_0} \ll 2\pi\;,$$

or

$$\frac{x^2+y^2}{2\lambda R_0} \ll 1\;. \tag{9}$$

Relation (9) is called the far-field approximation relation and is equivalent to assuming that the phase variation across the aperture is a linear function of the position. Within this approximation and by defining two spatial frequencies in the X and Y directions

$$\omega_x = -k\frac{X}{R_0} = -\frac{2\pi X}{\lambda R_0}\;,$$

$$\omega_y = -k\frac{Y}{R_0} = -\frac{2\pi Y}{\lambda R_0}\;, \tag{10}$$

the diffraction integral (8) becomes a more recognisable integral

$$\varphi\,(\omega_x,\omega_y) = \frac{iA}{\lambda}\,\frac{e^{-ik\cdot \mathbf{R}_0}}{R_0}$$

$$x \iint_\Sigma\, f(x,y)\,\exp\left[-\,i\left(\omega_x\,x + \omega_y\,y\right)\right]$$

$$x\,\left(\frac{\cos(\mathbf{n},\,\mathbf{r}_{oi}) - \cos(\mathbf{n},\,\mathbf{r}_{so})}{2}\right)\,dxdy\,. \tag{11}$$

The diffraction of an object given by (11) is the spatial Fourier transform of $f(x,y)$. The amplitude transmission of the object $f(x,y)$ may thus be interpreted as the superposition of mutually coherent plane-waves leaving the diffracting (object) screen, in the directions given by ( $(X/R_0)$, $(Y/R_0)$ ), and the Fraunhofer diffraction pattern is the interference pattern produced by a collection of plane-waves. The main consequence of interest for the projection microscope is that the intensity distribution of Fraunhofer diffractograms bears little or no resemblance with the aperture mask, i.e. the object geometry in the real space, As an example, the Fraunhofer diffraction from two complementary apertures is equal except for a constant phase and a bias at the origin (Babinet's principle). Physically, this means that the diffraction intensity distributions of complementary apertures will be identical with only a small difference in their brightness.

Since for a projection microscope working under Fraunhofer conditions, the projected diffractogram bears no resemblance with the shadow of the object projected on the screen, treatment of the diffractograms are needed to recover the object in the real pace by holography, for example, as discussed below.

## 4.2. FRESNEL DIFFRACTION

In Fraunhofer diffraction, the phase of the wave in the aperture is assumed to vary linearly across the whole aperture because we have an incident plane wave. In Fresnel diffraction, the source is close to the object and this is equivalent to assuming that a spherical wave of amplitude A at position $(x_S, y_S)$ illuminates the object located at $Z'$ away, i.e. the assumption of linear phase assumption is replaced by a quadratic phase variation over the object $\Sigma$.

The relation (5) needed for the solution of this diffraction problem will be rewritten with the complex transmission function $f(x,y)$ of the aperture, as

$$\varphi\,(P_i) = \frac{iA}{\lambda}\iint_\Sigma f(x,y)\,\frac{e^{-ik\cdot(\mathbf{r}_{oi} + \mathbf{r}_{so})}}{r_{oi}\,r_{so}}$$

$$x\,\left(\frac{\cos(\mathbf{n},\,\mathbf{r}_{oi}) - \cos(\mathbf{n},\,\mathbf{r}_{so})}{2}\right)dxdy\,. \tag{12}$$

This Huygens-Fresnel integral is non-zero only when the phase of the integrand is stationary. This is equivalent to stating that only the wave in a small aperture $\sigma$ around a stationary point $S_0$ will contribute to $\varphi(P_i)$. Physically, this means that only part of the wave propagating over paths nearly equal to the path predicted by geometrical optics using Fermat's principle will contribute to the diffraction. The stationary point $S_0$ $(x_0, y_0)$ is defined as the intersection point of the connecting line between the source and the observation point with the aperture plane, with co-ordinates

$$x_0 = \frac{Z'X + Zx_S}{Z' + Z} \quad , \quad y_0 = \frac{Z'Y + Zy_S}{Z + Z'} . \tag{13}$$

The parameters (13) can be used to express the spatial dependence of the phase in (12). Assuming a constant value for the obliquity factor across $\sigma$, introducing D, the distance between the source and a point $P_i$, and $\rho$ as:

$$D = Z + Z' + \frac{(X - x_S)^2 + (Y - y_S)^2}{2(Z + Z')} \quad ,$$

$$\rho = \frac{1}{Z} + \frac{1}{Z'} , \tag{14}$$

the Huygens-Fresnel integral becomes

$$\varphi(P_i) = \frac{iA}{\lambda\rho} \frac{e^{-ikD}}{D}$$

$$\times \iint_\sigma f(x,y) \exp\left( -\frac{ik}{2\rho} [ (x-x_0)^2 + (y - y_0)^2 ] \right) dxdy . \tag{15}$$

Relation (15) means that at any point $P_i$ on the observation screen a spherical wave, $e^{-ikD}/D$, originating at the source a distance D away is observed if no obstruction is present. In the presence of an obstruction, the amplitude and the phase of the spherical wave are modified by the integral in (15), this modification is called near-field diffraction. Physically, Fresnel diffraction means that the intensity distribution at the observation point is due only to wavelets from a small region $\sigma$ around $S_0$.

From the point of view of projection microscopy, the main conclusion is that the intensity distribution of Fresnel diffractograms contain intrinsically the shadows in the real space of the diffracting objects, conversely to the Fraunhofer diffraction images which are essentially in the Fourier space.

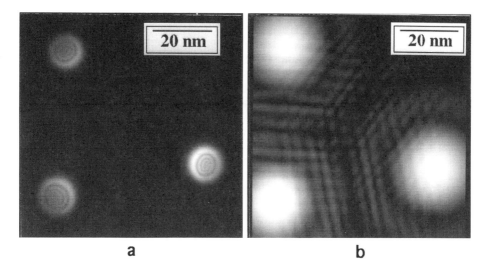

Figure 4. Numerical simulations of the diffractogrammes from a set of three 10 nm holes.
a) Fresnel diffraction, the interference fringes surrounded the hole geometry.
b) Fraunhofer diffraction, interference patterns are observable between the holes

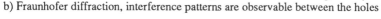

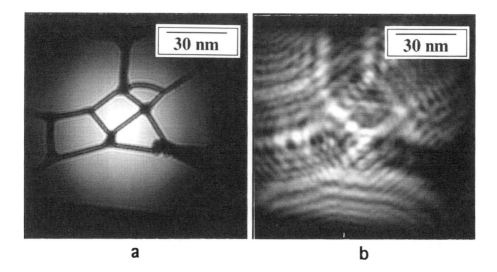

Figure 5: Numerical simulations of the diffractogrammes from a net of nano-fibres.
a) Fresnel diffraction, the interference pattern is composed of the shadow of the net
surrounded by interference fringes,
b) Fraunhofer diffraction, the interference pattern is in the Fourier space.

### 4.3. FRESNEL VS. FRAUNHOFER DIFFRACTOGRAMS

The best way to appreciate the differences between Fresnel and Fraunhofer diffraction is to compare the different diffractograms obtained from the same object in each of these two cases respectively and also the progressive modification of the image versus the source-object distance. Two examples of 2D masks were chosen, a set of three holes and a net of fibres. For each of these samples, Fraunhofer and Fresnel diffraction calculations were performed. The results show that under Fresnel diffraction the diffractograms are composed of the cast shadows of the objects surrounded with interference fringes, as consequence of (15). The position and the size of the holes (Fig. 4a) can be deduced directly from the Fresnel diffractograms if an absolute scaling in the object plane is known. The same remarks are valid for the net of fibres (Fig. 5a). Conversely, Fraunhofer diffractograms appear with intense interference patterns far away from the cast shadow of the object (Figs. 4b and 5b). In the case of the fibre network, it is difficult from the Fraunhofer diffractograms to perceive the mask of the object. For the three hole mask for example, the presence of regular dots between the big holes could be mis-interpreted as the casting shadow of smaller objects disposed regularly [11]. This can lead to serious mis-understandings as discussed in ref. [12] if distinction between the two image formation principles are not defined clearly before any analytical treatment.

### 5. Discussions

Among the questions that arise from the FPM experimental observations presented above, some are now open to discussion within the electron optics approach and diffraction theory.

### 5.1. FRESNEL DIFFRACTION FOR FPM

Fig. 6 shows the diffraction pattern obtained in the FPM with a carbon fibre and a comparison with a calculated Fresnel diffractogram. The calculated Fresnel diffraction pattern is obtained with the following parameters: $\lambda = 0.7$ Å, diameter of the wire = 14 Å and point source - object distance = 280 Å. Note that this last value corresponds to the distance of the virtual source to the object and it does not coincide with the actual distance of the nanotip apex with the object. The observations of nanometric details present along the fibres and the similarity between experimental and calculated Fresnel diffraction patterns indicate that the nanotips used were nearly ideal coherent point projectors and that the projection microscope with nanotips as electron sources operates in near field diffraction conditions.

Fig. 7 is the FPM diffraction image of two holes and a zoomed image of one of these two holes showing the Fresnel fringes inside it. The absence of interference fringes between the two holes in this image is an other experimental confirmation of the relation (15), i.e. Fresnel behaviour of this projection microscope.

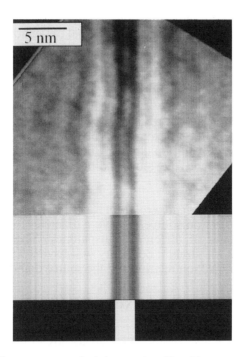

*Figure 6*: Fresnel diffractogrammes of a 1.4 nm carbon fibre. The upper image, obtained with the FPM, is in good agreement with the Fresnel numerical simulation pattern (the fibre mask is represented at the bottom of the image). Notice the little bending of the carbon fibre.

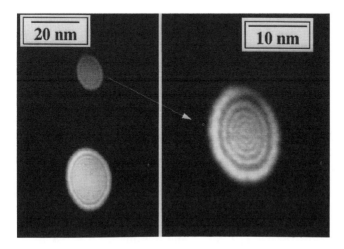

*Figure 7*: FPM image of a set of two holes. No interference pattern is observed outside the holes and the interference fringes inside the holes are Fresnel diffractogrammes.

This is consistent with our simple approach given in the upper paragraph and it is the justification of the name Fresnel projection microscope given to a projection microscope using a nanotip as electron field emission atom source.

## 5.2. LIMITING FACTORS FOR RESOLUTION

The limiting factors for the resolution in FPM are due mainly to the size of projection source, the presence of diffraction fringes and the image formation process. Additional image blurring, due to either the e-beam energy dispersion and the surrounding magnetic stray field, are also considered.

### 5.2.1. *Limit of geometrical resolution*
Consider the projection source to have a diameter of $2\rho$ and the boundary between two zones of different transparency. The overlap casting area corresponding to the straight boundary will have a width MN, whose value is given by the projection relation and the source diameter,

$$MN = 2\rho \frac{Z}{Z'} . \qquad (16)$$

At the object plane, this value corresponds to the limit of geometrical resolution:

$$\Delta_g = MN \frac{1}{M} \approx 2\rho , \qquad (17)$$

where M is the magnification factor given by (1). (17) means that $\Delta_g$ is of the order of the diameter $2\rho$ of the projection source.

This last value can be experimentally determined by the number of observable fringes present in the diffraction images. Two adjacent fringes are observable when the overlap of cast areas given by (16) is less than half the interfringe, that is

$$2\rho \frac{Z}{Z'} << \frac{1}{2} ( X_{n+1} - X_{n-1} ), \qquad (18)$$

where $X_n$ are the distances of the fringe maxima, or minima, to the position of the casting edge given by the geometrical projection. They are, at first approximation, given by [13]:

$$X_n^2 = \frac{\lambda Z (Z + Z')}{Z'} ( n - \frac{1}{4} ), \qquad (19)$$

with maxima for n = 1, 3, 5, ..., and minima for n = 2, 4, 6, ...

Taking into account (19) and for values of n not to near to 1, (18) can be rewritten as

$$2\rho \ll 0.5 \; (\frac{\lambda \, Z \, (Z + Z')}{Z'})^{1/2} \; \frac{1}{n^{1/2}}. \tag{20}$$

The dimension of the field emitting area from the nanotips could then be determined within the FPM experimental conditions ($Z' \approx 50$ nm, $Z \approx 10$ cm, $\lambda \approx 0.07$ nm) and counting the number of observable fringes in the FPM images, which is around 5. The experimental value is:

$$2\rho < 0.5 \; \text{nm} .$$

This value is in agreement with the known geometry of the nanotips [3], and it corresponds to the geometrical coherence of the field emission source area. The limit of the geometrical resolution in the FPM is then:

$$\Delta_g \approx 2\rho \; < 0.5 \; \text{nm} .$$

### 5.2.2. Resolution limit by diffraction
In presence of diffraction fringes, the localisation of the geometrical projected frontier can be achieved within a distance equal to half the distance of the first bright fringe :

$$X_1 \approx Z \, (\frac{\lambda}{Z'})^{1/2} . \tag{21}$$

This distance, taken back at the object plane, defines the resolution limit due to the diffraction, which is:

$$\Delta_d = \frac{1}{2} \frac{X_1}{M} \approx \frac{1}{2} (\lambda \, Z')^{1/2} \approx \frac{1}{2} (\lambda \, d)^{1/2} . \tag{22}$$

Within the experimental conditions of FPM, $\Delta_d$ varies in the range between 0.4 nm to 1 nm for source-object distance varying from 10 nm to 50 nm.

### 5.2.3. Resolution limit for statistical image acquisition
The theoretical limit for the visual detectability of small objects in a statistical noisy image is given by the Rose equation [14], which for the purpose of electron microscopy is

$$\Delta_a \geq \frac{5}{C \, (f N)^{1/2}} , \tag{23}$$

where $\Delta_a$ is the characteristic object size, C the contrast factor relative to the immediate surroundings (in our case we assumed to be 1), f is the efficiency of "electron utilisation" (we assumed to be 1), and N the number of incident electrons per unit area.

FPM images were taken with an exposure time of the order of 1/2 second, so $\Delta_a$ is in the range of 0.2 - 0.3 nm.

### 5.2.4. *Magnetic stray field and energy distribution blurring*

The measured permanent magnetic field is about 0.5 G ( ~0.5 × $10^{-4}$ Tesla) with ac stray field B($\omega$) in the range of 1 to 5 mG ( ~ $10^{-7}$ - 5 × $10^{-7}$ Tesla) nearby the FPM chamber. Under our experimental conditions, simple calculations [15] of the deviations of the image at the observation screen by Aharonov-Bohm and Lorentz force effects due to the stray fields give:

$$\Delta(i) \approx 2 \ 10^2 \times B \qquad (24)$$

with $\Delta(i)$ in meter and B in Tesla. For the measured range of the stray field B($\omega$) the deviations are from 20 to 100 $\mu$m. By taking into account the magnification factor M, this value in the object plane is less than 0.1 nm which defines the resolution limit $\Delta_B$ due to the magnetic stray field. It is substantially smaller than the other limits of resolution.

The fringe blurring $\Delta X_n$ can be also the consequence of the energy distribution $\Delta E$ of the incoming electrons

$$\frac{\Delta X_n}{X_n} \approx \frac{1}{4} \frac{\Delta E}{V}. \qquad (25)$$

Within the experimental values of the FPM and taking an upper limit of $\Delta E \approx 0.2$ eV, the fringe blurring due to the e-beam energy distribution:

$$\frac{\Delta X_n}{X_n} < 1 \ \text{to} \ 2 \ 10^{-4} \ ,$$

that means a resolution limit $\Delta_E$ negligible .

The different resolution limits are presented in table 1, it shows that the main limiting factors are coming from the source size, the diffraction and the image acquisition processes.

Table 1. Resolution limits due to different factors in the FPM

| Factors | Source size, $\Delta_g$ | Diffraction, $\Delta_d$ | Image process, $\Delta_a$ | Stray fields, $\Delta_B$ | Energy distr. $\Delta_E$ |
|---|---|---|---|---|---|
| Resolution limit in nm | < 0.5 | 0.4 to 1 | 0.2 to 0.3 | < 0.1 | negligible |

## 5.5. HOLOGRAPHY OR NOT HOLOGRAPHY

A discussion about the projection microscope cannot avoid the consideration of the notion of holography proposed by Garbor [16] in order to overcome diffraction effects intrinsic to electron microscopy and the inability of optical detectors to characterise the optical field which engenders the lost of the phase of the signal wave that is diffracted by the object. To surmount these shortcomings, a coherent reference wave provides the bias needed for recording both the amplitude and the phase of the signal wave, The signal and the reference waves interfere at the observation plane, named hologram plane. The hologram comes out as the record of the spatial intensity distribution of the resulting interference. It is then used to recover the object in the real space.

Let us consider the main points of this method which allows the obtaining of the real image of the object from its diffractogram. It is essentially a two step method: the hologram formation with an e-beam having a wavelength of $\lambda_0$, then the reconstruction from the magnified image of the hologram, by a factor $\lambda_1/\lambda_0$, with a photon beam having a wavelength $\lambda_1$.

In the first step, the hologram is the diffraction image recorded on a sensitive layer whose complex amplitude is, for the sake of simplicity, given as:

$$u \, e^{i\phi} = u_0 e^{i\phi_0} + u_1 e^{i\phi_1} \tag{26}$$

where $u_0 e^{i\phi_0}$ is from the direct wave, without the object, and $u_1 e^{j\phi_1}$ the diffracting term. The transparency T of the hologram is then given by:

$$T = K \, A^\Gamma , \tag{27}$$

where K and $\Gamma$ are constant values depending of the recording layer, and A the amplitude at the recording layer of the incoming wave given by (26):

$$A = [ \, u_0^2 + u_1^2 + 2 \, u_0 u_1 \cos (\phi_1 - \phi_0) \, ]^{1/2} . \tag{28}$$

During the second step, the transmitted wave u' obtained from the illumination of this hologram with an optical wave identical to the direct wave $u_0 e^{i\phi_0}$ (which is assumption 1) is:

$$u' = K[ \, u_0^2 + u_1^2 + 2 \, u_0 u_1 \cos (\phi_1 - \phi_0) \, ]^{\Gamma/2} u_0 e^{i\phi_0} . \tag{29}$$

Relation (29) can be simplified, by considering adequate values for K and $\Gamma$ and for $u_1 \ll u_0$ (which is assumption 2), to:

$$u' \sim e^{i\phi_0} [ \, u_0 + u_1 e^{+i (\phi_1 - \phi_0)} + u_1 e^{-i (\phi_1 - \phi_0)} \, ] . \tag{30}$$

Within these assumptions, the transmitted wave u' given by (30) differs from (26) only by the term $u_1 e^{-i(\phi_1 - \phi_0)}$ which is defined by Garbor as coming from a virtual twin object. The observed image through the hologram is then the superposition of the image of the object with the defocused image of the virtual twin object. This illustrates the real interest for holography, i.e. an easy recovery of the real images from the diffractograms.

From the practical point of view, the two assumptions put out in order to obtain (30) physically mean that:
- the reconstruction wave must be "identical" to the reference wave,
- and the amplitude of the diffracted wave $u_1$ must be negligible compared to the reference wave $u_0$.

Let us now look at these assumptions within the FPM experimental conditions and their practical consequences.

(1) To be identical to the reference wave, the reconstruction wave must first be a wave similar to the one emitted from the nanotips and secondly must have a value of $\lambda_1 = M \lambda_0$. For typical values of $\lambda_0 \approx 0.1$ nm and M in the range of $10^5$ to $10^6$, $\lambda_1$ must then have a value between $10^{-3}$ to $10^{-2}$ cm. Notwithstanding the difficulty to know the exact form of the emitting wave from the nanotips and to reproduce it from an optical source, the high values of $\lambda_1$ is a severe drawback for a rapid optical recovery of the object from the diffractogram.

(2) The interest of FPM is that small tip-object distances can be achieved in order to have a great magnification and to minimise the loss of resolution due to diffraction as estimated in (22). However, under such conditions the field of view covered by the object across the whole beam is a large part of the total beam and thus assumption 2 cannot be satisfied. For Fresnel projection microscopy under such a geometry, it is more realistic to speak about diffraction properties, instead of putting the stress on the concept of on-line holography [11], even if the original Garbor's scheme could suggest it.

(3) To overcome the shortcoming discussed just above, the use of a Fresnel bi-prism [17] inside the projection microscope has to be envisaged to perform holography, in order to preserve an exploitable reference beam.

(4) Holography is a very versatile technique to recover the real object from diffractograms in the Fourier space when the incident wave is a plane wave, as in most of transmission electron microscopy studies [18], even if numerical approaches do allow now phase retrieval directly from the diffractogramme intensity data [19]. For FPM, in which the diffractograms still keep the cast shadow of the object, holography technique loses most of its interest and the balance between the holographic process and direct numerical reconstruction from the Fresnel diffractograms is perhaps not in favour of holography.


## 6. Conclusions

Projection microscopy, i.e. formation of the shadow images cast on a screen by opaque objects illuminated by a radial source, was among the first techniques proposed for electron microscopy [20,8] and was the seed for diffraction microscopy. To enhance the resolution, the dimension of the source has been reduced mainly by focusing processes leading to two restrictive consequences: formation of diffraction fringes and aberrations from the electrostatic and magnetic lenses. To overcome the shortcomings due to the diffraction and aberrations, in particular those intrinsic to Fraunhofer diffraction process and to intensity recording of the diffractograms in the Fourier space, holography with the Fresnel bi-prism have been used with success in transmission electron microscopy. Recently, the controlled fabrication of nanotips and its direct use as radial atom source of low-energy coherent electrons make the Morton and Ramberg's projection microscope more appealing, this gives rise to the Fresnel projection microscope. The shadow images cast on the screen are in the real space, with important magnification ($> 10^6$) and without the aberrations intrinsic to the addition of electron lenses. The resolution is limited essentially to the source size and the Fresnel diffraction fringes. It is estimated to 0.5 nm for working voltages in the range of 200 V. The Fresnel projection microscope comes out essentially as a lensless high resolution low energy electron microscope. This is confirmed by the experimental observation of nanometric fibres (carbon, polymers, RNA) which gives directly images with details less than 1 nm in the real space.

*Acknowledgements:* The contributions of the Service Central d' Analyse / Département Instrumentation (CNRS-Solaize) are highly appreciated. It is a pleasure to thank Dr. S.T. Purcell and Mr. F. Feschet for fruitful discussions. This work has been supported by European Community Contracts (HCM and BRITE), by French and Spanish Government Agencies. One of us, V.S., is a CIFRE of La Lyonnaise des Eaux.

# REFERENCES

1. Vu Thien Binh, V. Semet and N. Garcia, Appl. Phys. Letters, **65**, 2493 (1994).
2. Vu Thien Binh, Journal of Microscopy **151,** 355 (1988); Vu Thien Binh and N. Garcia, J. de Phys. I1, **605** (1991); Vu Thien Binh and N. Garcia, Ultramicroscopy, **42-44**, 80 (1992)
3. For a review of nanotips see Vu Thien Binh, N. Garcia and S.T. Purcell, *Electron field emission from atom-sources: fabrication, properties and applications of nanotips*, in Advances in Imaging and Electron Physics, Vol. **95** (1995), Acad. Press, USA.
4. Vu Thien Binh, V. Semet and N. Garcia, Ultramicroscopy **58**, 307 (1995).
5. Vu Thien Binh, L. Bitar, V. Semet, N. Garcia and E. Taillandier, submitted

6.  A compilation can be found in M.P. Seah and W.A. Dench, Surface and Interface Analysis **1**, 2 (1979).

7.  Ch. Fert *Interférences, diffraction en optique électronique et leurs applications à la microscopie in Traité de Microscopie Électronique*, Vol. 1, pp. 333-390, C. Magnan Ed., Hermann, 1961, Paris, France; Joseph W. Goodman, in *Introduction to Fourier Optics* , McGraw-Hill Physical and Quantum Electronics Series, 1968, McGraw-Hill, N.Y., USA; J.M. Cowley in *Diffraction Physics*, North-Holland Publ. Co, 1981, Netherlands; P.W. Hawkes and E. Kasper in *Principles of Electron Optics, Wave Optics*, Vol. 3, Acad. Press, 1994, London, England.

8.  G.A. Morton and E.G. Ramberg, Phys. Rev. **56**, 705 (1939)

9.  For a review of STM techniques see C. Julian Chen, *Introduction to Scanning Tunneling Microscopy* , Oxford Series in Optical and Imaging Sciences, Oxford Univ. Press, New York USA, (1993).

10. Vu Thien Binh, S.T. Purcell, N. Garcia and J. Doglioni, Phys. Rev. Letters, **69**, 2527 (1992).

11. H.J. Kreuzer, K. Nakamura, A. Wierzbicki, H.W. Fink and H. Schmid, Ultramicroscopy **45**, 381 (1992); J.C.H. Spence and W. Qian, Phys. Rev. B **45**, 10271 (1993).

12. G.M. Shedd, J. Vac. Sci. Technol. A 12, 2595 (1994)

13. G. Bruhat, *Optique*, Masson et Cie, (1954), Paris, France.

14. A. Rose, Adv. Electronics **1**, 131 (1948); *Vision: Human and Electronics* Plenum Press, N.Y. (1973); R.M. Glaser, *Introduction to Analytical Electron Microscopy*, Plenum Press (1979) N.Y. pp.423.

15. R.P. Feynman, R.B. Leghton and M. Sands, *The Feynman Lectures in Physics* , Addison Wesley Pub. Co. London, Vol. II (1964).

16. D. Garbor, Holography, 1948-1971, Proc. IEEE, **60**, 655-668 (1972)

17. G. Möllenstedt and H. Druker, Natürwiss., **42**, 41-42 (1954); J. Faget and Ch. Fert, C.R. Acad. Sc. **243**, 2028-2029 (1956).

18. A. Tonomura, Rev. Mod. Phys. **59**, 639-669 (1987) and references inside.

19. J.R. Fienup, Appl. Optics, **21**, 2758 (1982).

20. H. Boersch, Z. Tech. Phys.,**20**, 346-350, (1939); Natürwiss., **28**, 709-711 (1940); ibid 711-712 (1940).

# AUTHOR INDEX

Bainier, C............................................105
Baida, F.............................................105
Berndt, R............................................175
Bitar, L.............................................277
Boujou, X.............................................95
Bozhevolnyi, S.I.....................................163
Carminati, R...........................................1
Castiaux, A...........................................95
Correia, A...........................................181
Courjon, D...........................................105
Dazzi, A.............................................119
De Fornel, F.........................................119
Dereux,A..............................................95
Dubreuil, N..........................................191
Durkan, C............................................257
Fischer, U.C.........................................247
Fuchs, H.............................................247
Fujihira, M..........................................205
Garcia, N...................................27,181,277
Girard, C.............................................95
Greffet, J.J...........................................1
Hare, J..............................................191
Haroche, S...........................................191
Hecht, B.............................................151
Hvam, J.M............................................263
Keilmann, F..........................................235
Keller, O.............................................63
Knight, J.C..........................................191
Koglin, J............................................247
Lefevre-Seguin, V....................................191
Madrazo, A............................................27
Madsen, S............................................263
Maradudin, A.A........................................41
Martin, Y............................................131
Massanell, J.........................................181
Mendez, E.............................................41
Mendoza-Suarez, A.....................................41
Moers, M.H.P.........................................223
Nieto-Vesperinas, M...............................27,41
Noordman, O.F.J......................................223
Novotny, L...........................................151
Olesen, T............................................263
Pic, C...............................................119
Pohl, D.W............................................151
Przeslawski, J.......................................181
Rahmani, A...........................................119
Raimond, J.M.........................................191
Ruiter, A.G.T........................................223
Salomon, L...........................................119
Segrink, F.B.........................................223

298

Semet, V.............................................277
Sharonov, M.........................................181
Shvets, I.V.........................................257
Vaez-Iravani, M.....................................143
Van Hulst, N.F......................................223
Van Labeke, D.......................................105
Vigneron, J.P........................................95
Vohnsen, B..........................................163
Vu Thien Binh.......................................277
Weeber, J.C.........................................119
Wickramasinghe, H.K.................................131
Willemsen, O.H......................................223
Zayats, A.V.........................................163
Zenhausern, F.......................................131
Zlatkin, A..........................................181

# SUBJECT INDEX

allowed direction of propagation...................154
antenna tips.......................................241
atomic force microscopy (AFM)..................132,205
atomic resolution.........................28,131,177
bacteriorodopsin...................................253
bandwidth..........................................106
beam deflection....................................206
cantilever.........................................206
Capacitance Microscopy.............................244
cavity field.......................................99
cavity Quantum Electrodynamics.....................191
charge interaction.................................18
coaxial waveguide..................................239
coherence degree...................................119
coherent illumination..........................13,277
configurational resonance..........................85
contrast...........................................143
coupled dipole method..............................159
coupling efficiency................................156
cutoff frequency...............................107,239
degrees of freedom.................................106
dielectric microsphere.............................191
diffraction........................................281
diffraction limit..................................205
electron beam......................................277
emission-mode microscope...........................152
energy flow........................................35
evanescent wavefield...........................107,195
extinction theorem.................................29
far-field......................................44,285
ferroelectric surface..............................182
fiber-optic interferometer.........................154
fiber-tip..........................................201
field confinement..................................239
field enhancement..................................33
field mapping......................................196
field pattern......................................192
fluorescence.........................128,205,215,223,
fluorescence lifetime..............................222
Fluorescence-Lifetime-Imaging-Microscopy (FLIM)....223
forbidden transition...............................224
Fraunhofer diffraction.............................283
Fresnel diffraction................................285
Fresnel-Projection-Microscope (FPM)................277
force gradient.....................................189
friction force microscope (FFM)....................205
grating........................................110,267
hologram.......................................10,277,293
image (allowed and forbidden)......................158
image artifacts....................................160
image recognition.............8,11,43,46,108,120,135,143

300

impedance (boundary condition).....................49
incoherent illumination........................13,16,123
infrared tip.......................................238
interference fringes..............................287
inverse problem...............................17,45,106
focusing...........................................241
light source......................................205
local-field theory.................................64
localization (weak)...............................164
long range interaction.............................73
macromolecules.....................................280
magnetic stray field..............................292
metallicfilm..........................151,165,205,247
microcavity........................................197
microlithography...................................208
molecular limit....................................71
molecular medium...................................75
monolayer film................................205,232
multiple multipole method.........................111
nanoline...........................................183
nanowriting........................................182
near-field...........................2,7,42,105,152,168,
                                    194
organic electroluminiscence.......................215
organic layer......................................215
passive probe......................................3
phase contrast microscopy in the near-field........144
phase shift (scatterd field)..................132,138
photo-chemical modifications......................263
photon emission....................................175
photon map.........................................177
Photon-Scanning-Tunneling-Microscope (PSTM)......3,109,163
plasmons (surface)......................32,52,151,163
plasmon resonance radiation........................7
point-dipole.......................................80
polarizing near-field microscopy..................145
power penetration depth...........................238
probe.........................................98,192,249
purple membrane...................................253
Q-factor......................................188,191
radiative rate....................................224
Radio-Wave Transmission Microscopy.................242
reflection-mode...................................257
resolution..........................27,103,105,147,179
                               ,181,205,235,243,247,277,290
resolution and coherence..........................126
resonant electromagnetic modes....................191
resonator (optical)................................95
Scanning-Electron-Microscope (SEM)................265
Scanning-Interferometric-Near-Field-Microscope (SIAM)...131
Scanning-Near-Field-Fluorescence-Microscope (SNFM)...211
Scanning-Near Field-Infrared-Microscopy (SNIM)......235
Scanning-Near-Field-Optical Microscope(SNOM).......1,108,
144,151,182,205,235,247,257,265

Scanning Surface Potential Microscope (SSPM)............205
Scanning-Tunnelling-Microscope (STM).........3,175,247,278
Scanning-Tunneling-Optical-Microscope (STOM)........95,108
scattering....................................1,2,29,42,152,164
shadow image.......................................295
shearforce.........................................108,185,206,265
shift (Goos-Hanchen).............................27,34
short range interaction............................73
spatial confinement................................84
speckle............................................17,120
standing wave......................................199
statistical image acquisition......................291
STM current........................................177
superresolution....................................1,106
surface defect.....................................41,96
system response....................................226
tetrahedral tip....................................247
Time Correlated Single Photon Counting............225
time window........................................232
time decay.........................................223,228
tip...............................................1,31,100,138,168,187,
                                                  192,208,248
TIP modes..........................................3, 176
tip-surface coupling...............................103
total internal reflection..........................7,28, 191
topography.........................................43,108,111,135,
                                                  168,178,183,208,268
transfer function..................................3,11,42
transmission electron microscope...................205
tunnel near-field optical microscope (TNOM)........151
tunnel photon density..............................157
vibrational frequency..............................237
waveguide..........................................51,239
whispering-gallery modes...........................191